AF892630

WATER STRESS IN PLANTS

Edited by **Ismail Md. Mofizur Rahman, Zinnat Ara Begum** and **Hiroshi Hasegawa**

Water Stress in Plants

http://dx.doi.org/10.5772/61897
Edited by Ismail Md. Mofizur Rahman, Zinnat Ara Begum and Hiroshi Hasegawa

Contributors

Yajuan Zhu, Mustafa Yildiz, Gholamreza Naser, Sina Shabani, Jan Adamowski, Peyman Yousefi, Mostafa K. Sarmast, Mehmet Cetin, Nurcan Yigit, Hakan Sevik, Nur Kaya, Daniela Simina Stefan

CBS, Edition **2017**

Published by InTech

Janeza Trdine 9, 51000 Rijeka, Croatia

The moral rights of the editor(s) and the author(s) have been asserted.
All rights to the book as a whole are reserved by InTech. The book as a whole (compilation) cannot be reproduced, distributed or used for commercial or non-commercial purposes without InTech's written permission.
Enquiries concerning the use of the book should be directed to InTech's rights and permissions department (permissions@intechopen.com).
Violations are liable to prosecution under the governing Copyright Law.
© The Editor(s) and the Author(s) 2016

Individual chapters of this publication are distributed under the terms of the Creative Commons Attribution 3.0 Unported License which permits commercial use, distribution and reproduction of the individual chapters, provided the original author(s) and source publication are appropriately acknowledged. More details and guidelines concerning content reuse and adaptation can be found at http://www.intechopen.com/copyright-policy.html.

Notice
Statements and opinions expressed in the chapters are these of the individual contributors and not necessarily those of the editors or publisher. No responsibility is accepted for the accuracy of information contained in the published chapters. The publisher assumes no responsibility for any damage or injury to persons or property arising out of the use of any materials, instructions, methods or ideas contained in the book.

Publishing Process Manager Dajana Pemac

Technical Editor SPi Global

Cover Designer InTech Design Team

Additional hard copies can be obtained from orders@intechopen.com

Water Stress in Plants, Edited by Ismail Md. Mofizur Rahman, Zinnat Ara Begum and Hiroshi Hasegawa
p. cm.
Print ISBN 978-953-51-2620-1
Online ISBN 978-953-51-2621-8

INTECH

open science | open minds

Contents

Preface

Water availability and water-use efficiency have a decisive influence on plant evolution. Water stress, which comprises both drought and flooding stress, is the most prominent environmental factor that affects the growth and distribution of vegetation. The impact of water stress in plants is it induces several morphological, physiological, biochemical and molecular changes, which cause reduced yield of the crops. The plants, however, try to adapt to the stress conditions using biochemical and physiological interventions.

The edited compilation is an attempt to provide new insights into the mechanism and adaptation aspects of water stress in plants through a thoughtful mixture of viewpoints. The book chapters, heterogeneous in nature, were invited by the publisher, and the authors are responsible for the accuracy of their contributions. The book consists of both review-like studies and the results of some new researches and case studies.

The compiled book is expected to be a useful document for professionals and researchers working on water stress in plants. We extend our sincere appreciation to the authors, who are from different countries, for their contribution to the book. We thank the InTech for inviting us to be editors of this book. We would like to extend our special appreciation to the Publishing Process Manger, Ms. Dajana Pemac, for her superb support.

Ismail Md. Mofizur Rahman
Fukushima University, Japan

Zinnat Ara Begum
Southern University, Bangladesh

Hiroshi Hasegawa
Kanazawa University, Japan

Chapter 1

Water Stress Hinders *In Vitro* Regeneration of Plants

Mustafa Yildiz, Emine Selcen Darcin and
Ramazan Beyaz

Additional information is available at the end of the chapter

http://dx.doi.org/10.5772/64664

Abstract

Plants could be propagated vegetatively via small parts of living tissue called as 'explant' on growth mediums under sterile conditions. Plant cell has the ability of forming whole fertile plant which is called 'totipotency', under *in vitro* culture conditions. High-frequency shoot regeneration is one of the main aims of *in vitro* culture and it is a prerequisite to guarantee the success in transformation studies and in clonal propagation of plants. It is well known that growth regulators in culture medium and the type of explant affect *in vitro* regeneration frequency significantly. In this chapter, the importance of tissue water content on *in vitro* culture response is discussed. Increasing water content of the explant before culture initiation gives rise to increased regeneration capacity. On the other hand, increasing the tissue's osmotic pressure enables the explant to intake water, all solutes and growth regulators from the growth medium which results in high-frequency shoot regeneration. However, tissues with lack of water are usually not successful in regenerating a satisfactory amount of shoots. The effect of water deficiency on explant's regeneration capacity and the methods to overcome this problem are discussed in this chapter.

Keywords: Plant *in vitro* culture, regeneration capacity, water, stress, growth

1. Introduction

Plant tissue culture includes techniques to propagate plants via somatic cells by using small parts called as explant on artificial growth mediums under sterile conditions. Shoots and roots are regenerated from explants, and consequently, the whole fertile plants are reconstituted under certain cultural conditions. Plant tissue culture belongs to totipotency meaning that a whole plant can be reproduced from a single cell in growth medium. Obtaining high-frequen-

INTECH
open science | open minds

© 2016 The Author(s). Licensee InTech. This chapter is distributed under the terms of the Creative Commons Attribution License (http://creativecommons.org/licenses/by/3.0), which permits unrestricted use, distribution, and reproduction in any medium, provided the original work is properly cited.

cy shoot regeneration is one of the major objectives for tissue culture studies that is also a prerequisite for an efficient transformation system and a clonal production of plants with interesting flowers and fruits massively for ornamental aims.

Plant tissue culture techniques have certain advantages over traditional propagation methods. Via tissue culture methods, thousands of mature plants having desirable traits such as good flowers, fruits and odor can be produced in a short time; endangered species which cannot propagate in native environment can be cloned easily by vegetative parts; genetically identical plants can be produced with large quantities; genetically modified plants can be regenerated from cultured cells; production of disease-, pest- and pathogen-free plants increase the plant production; and plants having seed germination and growing problems can be easily produced.

Plant growth regulators as media components affect the shoot regeneration capacity of explants. Tissue culture studies have tried to determine correct combinations of auxins and cytokinins for high-frequency adventitious shoot regeneration for related genotype. However, determination of optimum levels of auxins and cytokinins in growth medium is not the only way of increasing shoot regeneration capacity. It is reported that regeneration capacity of explant could be increased by adjusting the concentration, temperature and application period of NaOCl solutions used for surface sterilization [1] and manipulating physical microenvironment by altering distances among explants cultured resulted in increased shoot regeneration capacity [2]. Recently, it is noted that water capacity of the tissue affects explant's regeneration capacity significantly [3–5].

Source of life is based on water on the earth. Living is limited in a large proportion of terrestrial ecosystems according to water availability. The water content in an actively growing plant can be as much as 95% of its live weight. Water is needed in a plant for photosynthesis. Carbon dioxide and oxygen which is required for photosynthesis cannot be used by plant if they are not soluble in water. For this reason, water is the main factor for plant's existence and growth. Mineral ions such as potassium (K^+), sugars (glucose and sucrose) and amino acids are dissolved in water.

The decrease in growth, yield and quality by water stress has been reported in field conditions [6,7]. Plant survival is guaranteed by germination and seedling establishment and they are very important phases of plant life. Germination ratio diminishes with decreasing external water potential and there is a critical value of water potential for each species below which germination will not occur [8].

This chapter is aimed to show the effects of water deficiency in tissue on shoot regeneration capacity of the explants cultured under *in vitro* conditions. Moreover, increasing shoot regeneration frequency of explant by enhancing water content of the tissue is another issue this chapter focused on. All the results given here were based on three research studies.

2. The effect of increasing tissue water content on *in vitro* regeneration

It was reported that tissue water content affected explant's shoot regeneration capacity significantly [3]. Yildiz and Ozgen [3] have conducted a study to evaluate the effect of tissue water content on regeneration capacity of hypocotyl explants of flax (*Linum usitatissimum* L.). In the study, water-treated and non–water-treated hypocotyl explants of three flax cultivars ('Madaras', '1186 Sel.' and 'Clarck') obtained from Northern Crop Science Laboratories, North Dakota, USA, were compared with regards to fresh and dry weights, shoot regeneration percentage, shoot number per explant, shoot length and total shoot number per Petri dish. Sterilized seeds were germinated on a basal medium containing the mineral salts and vitamins of Murashige and Skoog (MS) [9], 3% (w/v) sucrose and 0.7% (w/v) agar. Hypocotyl segments of 5 mm length were excised from 7-day-old seedlings. Some hypocotyls were submerged in sterile distilled water and shook gently for 20 min before they were placed on growth medium for regeneration, while the others were directly cultured on MS medium containing 1 mg l^{-1} 6-benzylaminopurine (BAP) and 0.02 mg l^{-1} naphthaleneacetic acid (NAA) to regenerate. It is clear according to the results that there were sharp and statistically significant differences in all cultivars between water-treated and non-water-treated tissues related with all the characters examined (**Figure 1**).

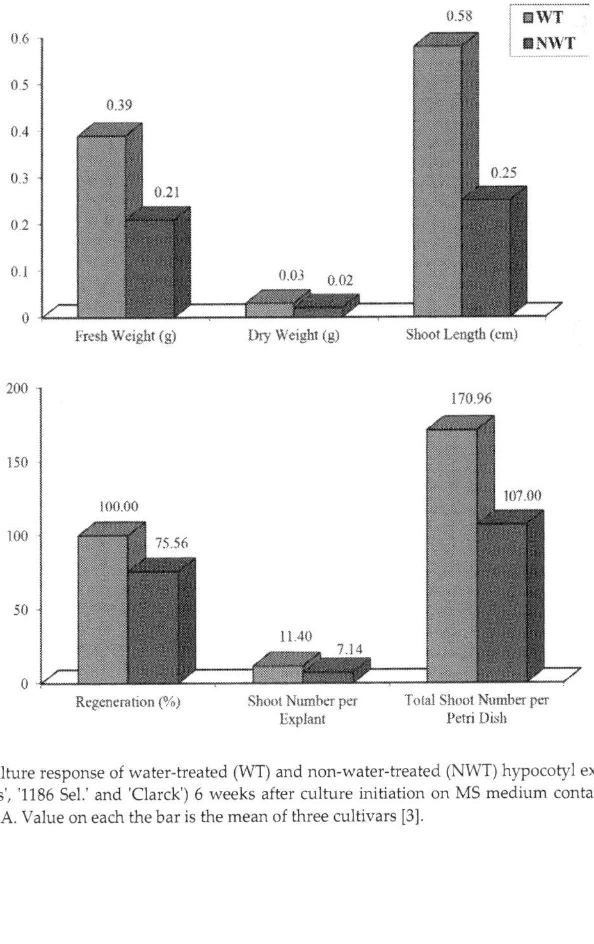

Figure 1. Tissue culture response of water-treated (WT) and non-water-treated (NWT) hypocotyl explants of three flax cultivars ('Madaras', '1186 Sel.' and 'Clarck') 6 weeks after culture initiation on MS medium containing 1 mg l^{-1} BAP and 0.02 mg l^{-1} NAA. Value on each the bar is the mean of three cultivars [3].

In the study, all explants were regenerated in water treatment application while only 75.56% of explants formed shoots in non-water treatment application. Water-treated explants had the highest fresh and dry weights compared to non-water-treated ones at the end of the culture (**Figure 2(a)** and **(b)**). Shoots grown from water-treated explants were more vital and well grown (**Figure 2(c)**) than the ones recovered from non-water-treated explants (**Figure 2(d)**). The highest shoot number per explant and total shoot number per Petri dish were obtained from the water-treated hypocotyl explants as 11.4 and 170.96, respectively. On the other hand, non-water-treated explants gave rise to only 7.14 shoots per explant and 107 shoots totally per Petri dish (**Figure 1**).

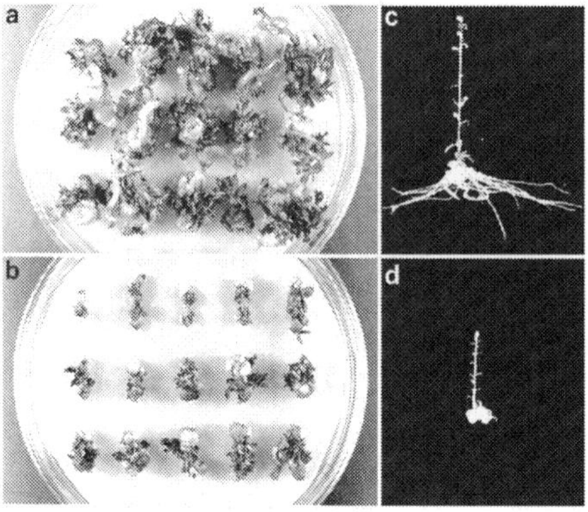

Figure 2. *In vitro* shoot regeneration in water-treated (a) and non-water-treated (b) hypocotyl explants of cv. '1886 Sel.'. *in vitro* root formation and plantlet development of shoots regenerated from water-treated (c) and non-water-treated (d) explants of cv. '1886 Sel.' [3].

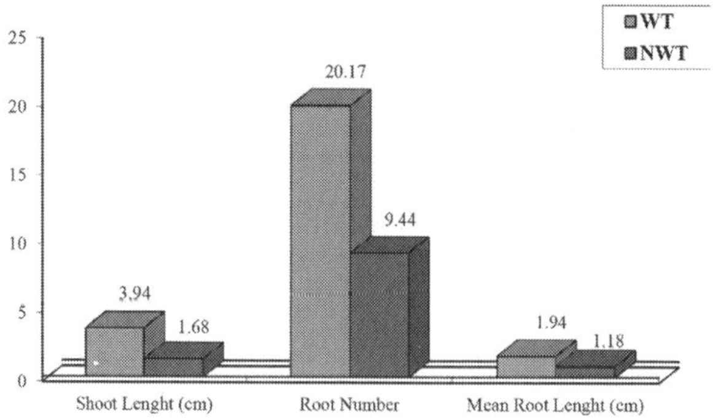

Figure 3. *In vitro* root development of shoots regenerated from water-treated (WT) and non-water-treated (NWT) hypocotyl explants of three flax cultivars ('Madaras', '1186 Sel.' and 'Clarck') on rooting medium enriched with 3 mg l^{-1} IBA 3 weeks after culture initiation. Value on each the bar is the mean of three cultivars [3].

Shoots got rooted on MS medium supplemented with indole-3-butyric acid (IBA) at a concentration of 3 mg l^{-1} for 3 weeks. The highest figures were recorded in the shoots regenerated from water-treated tissues (**Figures 2(c)** and **3**).

Statistically significant differences were observed in all parameters between the shoots which were regenerated from water-treated and non–water-treated explants. This sort of effects in water treatment got also noted in the rooting stage. It means that shoots which were regenerated from water-treated explants got more capable of establishing new plantlets than the ones which were grown from non–water-treated explants.

It could be concluded that the lower levels of all parameters of non–water-treated explants were directly due to a decreasing amount of water uptake from the environment and consequently, a reduced mobilization of plant growth regulators. Application of water treatment to explants before culture initiation enriched the tissue's water content and so enabled all solutes and plant growth regulators to transfer into the tissue, providing all cells with a high regeneration capacity and consequently, increasing explant's tissue culture response. Increased growth in water-treated explants was confirmed by Naylor's [10] study which stated that plant growth regulators promote cell division and cell elongation. It has also been reported that decreased germination and seedling growth in stressed rice seedlings was due to decreased mobilization of starch and α-amylase activity [11].

It is understood that pretreatment of explants with water before culture initiation increased the permeability of the epidermis layer and caused to high metabolic activity by increased uptake of water and hormone from the growth medium. Higher fresh and dry weights of water-treated hypocotyls at the end of culture could be attributed to an increase in the absorption of water and other components from the growth medium by means of high permeable epidermis membrane. Water-treated tissues were observed bigger in size than non–water-treated ones in all cultivars as reported by Dale [12], who pointed out that the fresh weight increase causes the cell enlargement with water absorption, cell vacuolation and turgor-driven wall expansion in this study. The increase in dry weight got closely related to cell division and new material synthesis [13]. Dry weight increase of water-treated tissues is caused by an increase in carbohydrate metabolism resulting from the increased water uptake. Besides, lower levels of all the parameters of non-water-treated tissues caused directly a decreased water uptake through the environment and nevertheless, a decreased mobilization of plant growth regulators. Inhibition of the cell division, elongation of cell, or both of them led to the inhibition of growth under water stress conditions [14]. Cell elongation is affected by osmotic water absorption. Osmotic stress lead to biochemical changes in cell wall during growth [15]. Osmotic stress inhibits water uptake which is vital for germination and growth [16]. And water stress affects the level of plant hormones significantly [17].

3. The effect of increased water absorption on shoot regeneration

In another study conducted by Yildiz et al. [4], hypocotyl explants of three flax cultivars ('Omega', 'Fakel' and 'Ariane'), which were pretreated and non-pretreated before culture, were cultured for regeneration. In the study, two regeneration methods, which were based on two different pretreatment applications, were compared with the conventional regeneration protocol in which explants were directly cultured on MS medium supplemented with 1 mg l^{-1} BAP and 0.02 mg l^{-1} NAA. Hypocotyl explants were kept in sterile cabin under air flow for 30 min in order to make them dry as reported by Christmann et al. [18] in the first and second pretreatment applications in order to decrease the tissue water content and to help the tissues gain the ability to uptake increased amount of water, all solutes and plant growth regulators from the growth medium via tissue's higher osmotic pressure. Later, explants were submerged in MS solution having 1 mg l^{-1} BAP and 0.02 mg l^{-1} NAA for 15 min in both pretreatment applications. Then, explants were cultured on MS medium without growth regulators in the first pretreatment application and on MS medium containing 1 mg l^{-1} BAP and 0.02 mg l^{-1} NAA in the second pretreatment application. It was thought that drying of explants under air flow in sterile cabin increased tissue's osmotic pressure and enabled all cells to absorb more growth regulators along with water in both pretreatment applications by immersing explants into liquid. On the other hand, explants were cultured on MS medium containing 1 mg l^{-1} BAP and 0.02 mg l^{-1} NAA only in the second pretreatment application that means tissues maintained uptaking water and growth regulators from the medium and this led to the higher results in all parameters studied as noted by Yildiz and Ozgen [3]. Okubo et al. [19] has reported that regeneration capacity was affected by endogenous hormone levels of tissue significantly. Fatima et al. [20] has also reported that plant growth is affected by the internal factors such as chemicals and mineral nutrients. Endogenous levels of growth regulators of the plant tissue determine the amount of exogenous plant growth regulators required for regeneration [20]. It was firstly reported that keeping the explants in sterile distilled water for a while before culture initiation promoted the regeneration capacity of explants by increasing tissue's water content and enabling water, all solutes and growth regulators to transfer into the tissue more easily [3].

In accordance with the results, there were statistically important differences among pretreated and non-pretreated hypocotyls in all cultivars. The highest results in all parameters studied were recorded from the second pretreatment application. On the other hand, the lowest results were obtained from the first pretreatment application in which explants were cultured on MS medium without growth regulators in all cultivars after submerging them in MS solution having 1 mg l^{-1} BAP and 0.02 mg l^{-1} NAA for 15 min (**Figure 4**).

Higher results in the fresh and dry weights could be attributed to higher metabolic activity caused by an increase in the absorption of water and growth regulators from the growth medium. From the results of the second pretreatment application, it might be easily seen that culturing explants on MS medium having 1 mg l^{-1} BAP and 0.02 mg l^{-1} NAA after submerging them in liquid MS medium having 1 mg l^{-1} BAP and 0.02 mg l^{-1} NAA increased the tissue's growth regulators' level leading to the higher fresh and dry weights. In fact, transferring explants on MS0 medium after treating them with liquid MS that has 1 mg l^{-1} BAP and 0.02 mg l^{-1} NAA for a moment in the first pretreatment application, growth regulators of tissues did not seem to be sufficient for high scores according to fresh and dry weights. Culturing explants directly on MS medium containing 1 mg l^{-1} BAP and 0.02 mg l^{-1} NAA were not enough again in the increasing tissue's growth regulators' content to obtain higher scores in characters examined in the non-pretreatment application. All the explants regener-

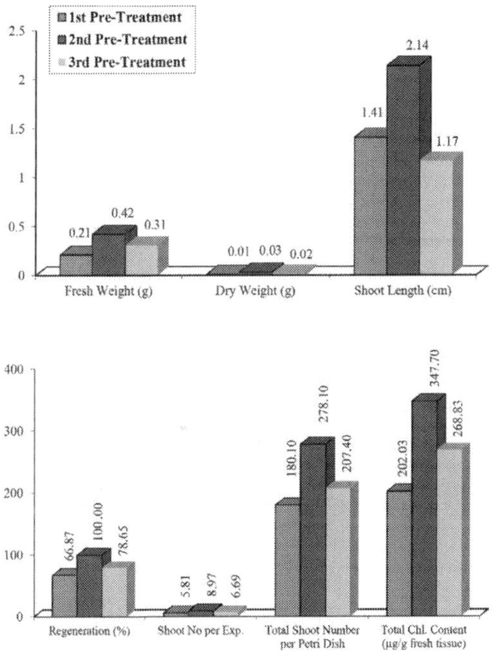

Figure 4. Tissue culture response of pretreated and non-pretreated hypocotyls of three flax cultivars ('Omega', 'Fakel' and 'Ariane') 6 weeks after culture initiation. Value on each the bar is the mean of three cultivars [4].

ated in the second pretreatment application successfully with the regeneration percentage of 100% (**Figures 4** and **5**).

The highest results in shoot number per hypocotyl and shoot length were obtained from second pretreatment application in all cultivars studied. The highest shoot number per hypocotyl was recorded as 8.97. The highest score related to shoot length was 2.14 cm. Shoot regeneration capacity of hypocotyls increased significantly in second pretreatment application. The explants to which second pretreatment application was carried out were more vital and well-grown and more capable of regeneration (**Figures 5(b)** and **6(b)**). The highest total shoot number per Petri dish was obtained as 278.10 from second pretreatment application. Total shoot number per Petri dish was reported as a good indicator of the success in both shoot regeneration percentage and shoot number per explant [21]. The highest result of the total chlorophyll content was achieved from the second pretreatment application as 347.70 µg/g fresh tissue. Emerson [22] reported that there exists a close relationship between photosynthesis and chlorophyll content. Chlorophyll content of leaf is thought as a sign of photosynthetic capacity of tissues [22–25] playing a critical role in plant growth and development [26] and its amount alters under stress conditions [27–29]. Gireesh [30] has informed that chlorophyll can be used for measuring growth.

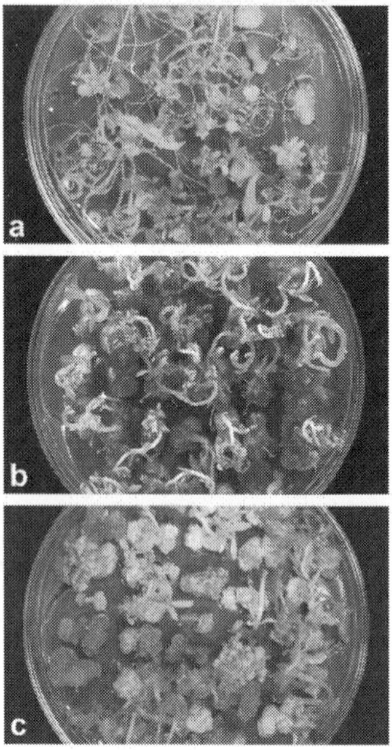

Figure 5. Shoot regeneration from hypocotyl explants of flax cv. 'Omega' [4]. (a) The first pretreatment application: hypocotyls dried for 30 min in sterile cabin and then they were imbibed to liquid MS medium containing 1 mg l^{-1} BAP and 0.02 mg l^{-1} NAA for 15 min, and consequently, cultured on MS medium without growth regulators, (b) the second pretreatment application: hypocotyls got dried by waiting for 30 min in sterile cabin and then were imbibed to liquid MS medium containing 1 mg l^{-1} BAP and 0.02 mg l^{-1} NAA for 15 min, and finally, cultured on MS medium having 1 mg l^{-1} BAP and 0.02 mg l^{-1} NAA and (c) non-pretreatment application: hypocotyl explants got directly cultured on MS medium containing 1 mg l^{-1} BAP and 0.02 mg l^{-1} NAA.

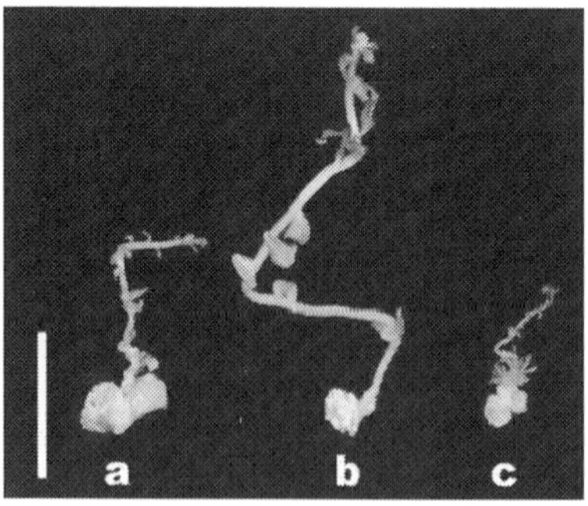

Figure 6. Regenerated shoots of cv. 'Omega' from (a) first pretreatment application, (b) second pretreatment application and (c) non-pretreatment application 6 weeks after culture initiation (bar = 1.0 cm) (original).

In accordance with the results, it might be concluded that the lower levels of all the parameters which were recorded in the first and third pretreatment applications caused from a decreased uptake of water and growth regulators directly from the medium. Higher shoot regeneration has been significantly affected by tissue water content [3]. Keeping explants in liquid medium containing 1 mg l^{-1} BAP and 0.02 mg l^{-1} NAA for a while before culture enabled water, all solutes and growth regulators to transfer through the tissue easily, providing all the cells a high regeneration capacity.

4. The effect of water deficiency originated stress in explant on shoot regeneration capacity

In the study conducted by Derelli et al. [5], the effects of water deficiency on shoot regeneration capacity of the explant were evaluated. Flax (*L. usitatissimum* L.) cv. 'Clarck' seeds, which were obtained from 'Northern Crop Science Laboratories', North Dakota, USA, got used in the study. Before germination, seeds were surface sterilized with 40% commercial bleach containing 5% sodium hypochlorite at 10°C for 12 min with continuous stirring and then were washed three to four times with sterile water at the same temperature [31]. Sterilized seeds were germinated on MS medium in Magenta vessels. All cultures were incubated at 24 ± 1°C with a 16-h light/8-h dark photoperiod. Hypocotyl explants were removed as in 5 mm in length from 10-day-old sterile seedlings. Some of the hypocotyls were directly transferred to regeneration medium, while some of them were kept in sterile cabin under air flow for 30 min to decrease the water content of the tissue. Hypocotyl explants were cultured on MS medium which contains 1 mg l^{-1} BAP and 0.02 mg l^{-1} NAA. Four weeks after culture initiation, the results obtained from two pretreatment applications were compared with respect to regeneration percentage, shoot number per explant, the highest shoot length per explant and total shoot number per Petri dish.

	Regeneration (%)		Shoot number per explant		The highest shoot length (cm)		Total shoot number per Petri dish	
	Non-dried	Dried	Non-dried	Dried	Non-dried	Dried	Non-dried	Dried
	100	100	4.85	4.10	3.32	2.46	48.50	41.00
t value	0.000^{ns}		2.585*		2.296*		2.585*	

^{ns}Not significant.
*Statistically significant at 0.05 level.

Table 1. The effect of water deficiency in explant on tissue culture response of flax (*Linum usitatissimum* L.) cv. 'Clarck'.

The highest results in the study were obtained from the treatment in which hypocotyl explants were directly transferred to regeneration medium without drying. On the other hand, keeping explants under air flow in sterile cabin for 30 min led to evaporation of more water from the explant and consequently, a decrease in the regeneration capacity (**Table 1**). For this reason, while working in the sterile cabin, explants are to be isolated and placed on growth medium as quickly as possible to protect the regeneration capacity thinking that air flow in the environment can have negative influence on the tissue.

There was no difference between non-dried and dried explants with respect to regeneration percentage. Explants from both treatments formed shoots. The highest results were recorded from non-dried explants, which had higher water content than dried ones, as 4.85, 3.32 cm and 48.50 in shoot number per explant, the highest shoot length and total shoot number per Petri dish, respectively. On the other hand, the lowest results were obtained from dried explants as 4.10, 2.46 cm and 41.00, respectively (**Table 1**).

Lower results from dried explants could be attributed to a decreased water potential of explant tissue and difficulty in distribution of all solutes and growth regulators among cells.

5. Conclusion

The purpose of tissue culture studies is to obtain high-frequency shoot regeneration that is also a prerequisite for an efficient transformation system and a clonal propagation of plants. The introduction of foreign genes which code agronomically important traits into plant cells has not got any meaning if transgenic plants are not recovered from the transformed cell(s). For this reason, tissues with high regeneration capacity should be used. Regeneration capacity of the tissue is the key factor affecting the success of transformation studies. Types, concentrations and combinations of plant growth regulators affect *in vitro* explant growth significantly. Correct concentrations and combinations of auxins and cytokinins should be determined to obtain high frequency adventitious shoot regeneration for related genotype. However, determining the explant type, and the correct concentrations and the combinations of growth regulators is not sufficient for the high frequency shoot regeneration. Shoot regeneration frequency can always be higher than the one we obtain in theory, as every cell has got an ability to form a whole fertile plant under *in vitro* conditions. Many factors affecting regeneration capacity of explant have not been found out yet. Such as, a recently reported technique that utilizes competition among the explants is quite effective to increase shoot regeneration capacity [2]. In this way, the unknown factors affecting regeneration capacity of explants ought to be determined to increase the success of tissue culture studies. In this chapter, the importance of water on shoot regeneration capacity as a main component of all living cells was discussed. Results of research studies given in this chapter showed that enriching tissue with water give rise to higher values with respect to tissue culture response. On the contrary, water deficiency in tissue decreased the regeneration capacity of explant significantly. From now on, water content of the explant should be considered as one of the most important factors such as growth regulators and explant type regarding higher tissue culture response.

http://dx.doi.org/10.5772/64664

Author details

Mustafa Yildiz[1*], Emine Selcen Darcin[2] and Ramazan Beyaz[3]

*Address all correspondence to: myildiz@ankara.edu.tr

1 Department of Field Crops, Faculty of Agriculture, Ankara University, Diskapi, Ankara, Turkey

2 Department of Field Crops, Faculty of Agriculture and Natural Sciences, Bilecik Seyh Edebali University, Bilecik, Turkey

3 Department of Soil Science and Plant Nutrition, Faculty of Agriculture, Ahi Evran University, Bagbasi, Kırsehir, Turkey

References

[1] Yildiz M, Ekiz H. The effect of sodium hypochlorite solutions on *in vitro* seedling growth and regeneration capacity of sainfoin (*Onobrychis viciifolia* Scop.) hypocotyl explants. Canadian Journal of Plant Science. 2014; 94: 1161–1164.

[2] Yıldız M. Evaluation of the effect of *in vitro* stress and competition on tissue culture response of flax. Biologia Plantarum. 2011; 55(3): 541–544.

[3] Yildiz M, Ozgen M. The effect of a submersion pretreatment on *in vitro* explant growth and shoot regeneration from hypocotyls of flax (*Linum usitatissimum*). Plant Cell Tissue and Organ Culture. 2004; 77: 111–115.

[4] Yildiz M, Ozcan S, Telci C, Day S, Özat H. The effect of drying and submersion pretreatment on adventitious shoot regeneration from hypocotyl explants of flax (*Linum usitatissimum* L.). Turkish Journal of Botany. 2010; 34: 323–328.

[5] Derelli E, Allahverdikhan Vaziri P, Mirzapour M, Yildiz, M. The effect of water deficiency originated stress in tissue on shoot regeneration of flax (*Linum usitatissimum* L.) hypocotyl explants. Research Journal of Agricultural Sciences. 2012; 5(1): 153–156.

[6] Fisher RA. Influence of water stress on crop yield in semi arid regions. In: Turner NC, Kramer P (eds.) Adaptation of Plants to Water and High Temperature Stress. Wiley and Sons, New York; 1980. pp. 323–340.

[7] Kriedeman PE, Barrs HD. Citrus orchards. In: Koziowski TT (ed.) Water Deficit and Plant Growth. Academic Press, New York; 1981. pp. 325–417.

[8] Hadas A. Water uptake and germination of Leguminous seeds under changing external water potential in osmotic solutions. Journal of Experimental Botany. 1976; 27(3): 480–489.

[9] Murashige T, Skoog F. A revised medium for rapid growth and bioassays with tobacco tissue cultures. Physiologia Plantarum. 1962; 15: 431–497.

[10] Naylor AW (1984) Hormonal regulation of development. II. The function of hormones from the level of the cell to whole plant, In: Scott TK (ed.) Encyclopedia of Plant Physiology, New Series. Springer Verlag, Berlin; 1984. pp. 180–185.

[11] Lin CC, Kao CH. NaCl stress in rice seedlings: starch mobilization and the influence of GA_3 on seedling growth. Botanical Bulletin Academia Sinica. 1995; 36: 169–173.

[12] Dale JE. The control of leaf expansion. Annual Review of Plant Physiology. 1988; 39: 267–295.

[13] Sunderland N. Cell division and expansion in the growth of the leaf. Journal of Experimental Botany. 1960; 11: 68–80.

[14] Hsiao TC. Plant responses to water stress. Annual Review of Plant Physiology. 1973; 24: 219–270.

[15] Van Volkenburg E, Boyer JS. Inhibitory effects of water deficit on maize leaf elongation. Plant Physiology. 1985; 77: 190–194.

[16] Kahn A. An analysis 'dark-osmotic inhibition' of germination of lettuce seeds. Plant Physiology. 1960; 35: 1–7.

[17] Morgan PW. Effects of abiotic stresses on plant hormone systems. In: Koziowski TT (ed.) Stress Responses in Plants: Adaptation and Acclimation Mechanism. Wiley-Liss, New York; 1990. pp. 113–146.

[18] Christmann A, Hoffman T, Teplova I, Grill E, Muller A. Generation of active pools for abscisic acid revealed by *in vivo* imaging of water-stressed *Arabidopsis*. Plant Physiology. 2005; 137: 209–219.

[19] Okubo H, Wada K, Uemoto S. *in vitro* morphogenetic response and distribution of endogenous plant hormones in hypocotyl segments of snapdragon (*Antirrhinum majus* L.). Plant Cell Reports. 1991; 10: 501–504.

[20] Fatıma Z, Mujib A, Fatıma S, Arshi A, Umar S. Callus induction, biomass growth, and plant regeneration in *Digitalis lanata* Ehrh: influence of plant growth regulators and carbonhydrates. Turkish Journal of Botany. 2009; 33: 393–405.

[21] Yıldız M, Saglik Ç, Telci C, Erkilic EG. The effect of *in vitro* competition on shoot regeneration from hypocotyl explants of *Linum usitatissimum*. Turkish Journal of Botany. 2011; 35: 211–218.

[22] Emerson R. Chlorophyll content and the rate of photosynthesis. Proceedings of the National Academy of Sciences of the United States of America. 1929; 15(3): 281–284.

[23] Pal RN, Laloraya MM. Effect of calcium levels on chlorophyll synthesis in peanut and linseed plants. Biochemie und Physiologie der Pflanezen. 1972; 163: 443–449.

[24] Wright GC, Nageswara RRC, Farquhar GD. Water use efficiency and carbon isotope discrimination in peanut under water deficit conditions. Crop Science. 1994; 34: 92–97.

[25] Nageswara RRC, Talwar HS, Wright GC. Rapid assessment of specific leaf area and leaf nitrogen in peanut (*Arachis hypogaea* L.) using chlorophyll meter. Journal of Agronomy and Crop Science. 2001; 189: 175–182.

[26] Yang X, Wang X, Wei M. Response of photosynthesis in the leaves of cucumber seedlings to light intensity and CO_2 concentration under nitrate stress. Turkish Journal of Botany. 2010; 34: 303–310.

[27] Rensburg LV, Kruger GHJ. Evaluation of components of oxidative stres metabolism for use in selection of drought tolerant cultivars of *Nicotiana tabacum* L. Journal of Plant Physiology. 1994; 143: 730–737.

[28] Kyparissis A, Petropoulou Y, Manetas Y. Summer survival of leaves in a soft-leaved shrub (*Phlomis fruticosa* L., Labiatae) under Mediterranean field conditions: Avoidance of photoinhibitory damage through decreased chlorophyll contents. Journal of Experimental Botany. 1995; 46: 1825–1831.

[29] Jagtap V, Bhargava S, Sterb P, Feierabend J. Comparative effect of water, heat and light stresses on photosynthetic reactions in *Sorghum bicolor* (L.) Moench. Journal of Experimental Botany. 1998; 49: 1715–1721.

[30] Gireesh R. Proximate composition, chlorophylla, and carotenoid content in *Dunaliella salina* (Dunal) Teod (Chlorophycea: Dunaliellaceae) cultured with cost-effective seaweed liquid fertilizer medium. Turkish Journal of Botany. 2009; 33: 21–26.

[31] Yildiz M, Er C. The effect of sodium hypochlorite solutions on *in vitro* seedling growth and shoot regeneration of flax (*Linum usitatissimum*). Naturwissenschaften. 2002; 89: 259–261.

Chapter 2

Water Stress Induced by Enrichment of Nutrient and Climate Change Factors

Daniela Simina Stefan and Mircea Stefan

Additional information is available at the end of the chapter

http://dx.doi.org/10.5772/64665

Abstract

Human activities accelerate the rate of water, air, and soil environment degradation. In this paper, we propose to present the effects of important stressors for water bodies, represented by continuous enrichment with nutrients and climate change. Nutrients' concentration increases especially nitrogen and phosphorus, associated with temperature increases and extreme weather events, involve important physical, chemical, and biological alterations of water quality. The effect of the stressors' factors can be seen on the main parameters which characterize water: temperature, pH, dissolved oxygen, transparency, chlorophyll *a*, nitrogen and phosphorus compounds, and plankton population. The main changes which occur in the water reservoirs consist of modifications of taste and odor, increased acidification, decreased transparency, oxygen depletion, increased toxicity, increased sediments quantity, excessive growth of phytoplankton, and macrophytes vegetation. The water quality of lakes, streams, and estuaries can be assessed using the trophic status that can be described mainly using limiting nutrients' concentrations (N total, P total), primary productivity (chlorophyll *a*), and Secchi disc parameters, and also the Carlson's index that includes all of these.

Keywords: eutrophication, stress factors, nitrogen, phosphorus, climate change, temperature, dissolved oxygen, transparency, chlorophyll a, trophic status assessment, Carlson's index

1. Introduction

Water has a unique place on the planet as it supports life on the earth. Clean water is an important resource for drinking, irrigation, industry, transportation, recreation, fishing, hunting, the biodiversity support, and sheer aesthetic enjoyment.

INTECH
open science | open minds

© 2016 The Author(s). Licensee InTech. This chapter is distributed under the terms of the Creative Commons Attribution License (http://creativecommons.org/licenses/by/3.0), which permits unrestricted use, distribution, and reproduction in any medium, provided the original work is properly cited.

The main water stressors are human activities reflected by nutrients' inputs and climate change by increasing temperature and extreme weather phenomena. Nutrients, especially phosphorus and nitrogen from various sources, and increasing temperature are the major causes of degradation of the aquatic ecosystems, namely eutrophication.

Eutrophication can be a natural process in surface waters occurring as they age through geological time, over hundreds or thousands of years or it can be very fast when the nutrients are present in high concentrations, due to anthropogenic activities and climate change [1, 2].

Degradation of these vital water resources (coastal areas, lakes, and reservoirs all over the world) can be measured by the loss of natural systems, followed by the modification of the trophic chains with their component species, and the increase in the number of individuals of a species in preference to others [3].

It is considered one of the major forms of water stress, which is extremely variable being influenced by the specific characteristics of sites such as nutrient stoichiometry, biodiversity, climate-related factors (temperature, precipitation, and storming), and the basin geomorphology [4, 5].

The main sources of nutrients (nitrogen and phosphorus), its effect on water quality associated with influence of climate change factors, are presented in this chapter. The stressors' effects, nutrients, and climate exchange were highlighted by the parameters: temperature, pH, Secchi disc (SD) transparency, chlorophyll *a* (CHL), dissolved oxygen (DO), total phosphorus (TP), total nitrogen (TN), and plankton populations. The trophic stage was assessed using temperature, limiting nutrients concentrations (N total, P total), and their ratio TN/TP, primary productivity (chlorophyll *a*), and transparency (Secchi disc) parameters, and also the Carlson's index that includes all of these. To illustrate how one can achieve a surface water quality evaluation was presented using the trophic status of Lake Snagov, Romania, assessment.

2. Water stressors

Human activities accelerate the degradation rate of water, air, and soil. Continuous enrichment with nutrients and climate change are important stressors for water bodies. Increasing nutrients' concentrations associated with increasing temperature and extreme weather events involve important changes in the physical, chemical, and the biological configuration of the waters' characteristics [6].

2.1. Nutrient inputs

2.1.1. Sources of nutrients

The mains stressors that influence the water quality and trophic chain reaction are macronutrients, such as phosphorus (P), nitrogen (N), silicon (Si), and micronutrients such as potassium (K), sulfur (S), calcium (Ca), magnesium (Mg), iron (Fe), and molybdenum (Mo) are also needed. N, P, and K are considered primary nutrients, and N and P are the major limiting nutrients in most aquatic environments [6–8].

The nutrient inputs in natural water body by both point and disperse sources. The point sources may be wastewater effluents (domestic and industrial)—most importantly, runoff and leach flows from waste disposal systems, infiltrations from animal feedlots, unsewered industrial sites runoffs, sanitary sewers overflows, runoff from constructions sites, erosion into the lake. The most important dispersed (nonpoint) sources are leachate of synthetic and natural fertilizes from agriculture parcels and forest, runoff and infiltration from animal feedlots, runoffs from agriculture/irrigation, pasture and range, urban runoff from not-sewered areas, septic tank leachates, and atmospheric deposition on water surface [1, 10].

The results of enrichment of nutrient consist in the increase of aquatic primary production and lead to visible algal blooms causing high turbidity and increasing anoxia in the deeper parts, thus increasing the acidity and the modified aquatic ecosystems [9–11]. All these involve the water quality deterioration, drinking water treatment problems, and decrease in the perceived aesthetic value of the water body. The physical and chemical properties of the water influence the distribution and trophic dynamics in the water body. Depending on the content of nutrient and the production of organic materials, the water can have a trophic (degradation) level lower or higher.

2.1.2. Water trophic level classification

According to the content of mineral nutrients, and the effect of these on primary production, the trophic level classifications of water can be characterized using the terms as follows [12–14]:

- *oligotrophic*, poor in nutrients (nitrogen, phosphor)—nutrient deficient, small production of organic matter—primary production (amount of organic carbon produced by photosynthesis), clear waters, well illuminated, well oxygenated, low algae production, diatoms predominant;

- *mesotrophic*, medium levels of nutrients, intermediate level of primary production, clear water, and ponds with beds of submerged aquatic plants;

- *eutrophic*, more nutrients, primary production, and higher organic compounds, the oxygen concentration decreases with depths and the deep water layer is anoxic during summer, development of microalgae and cyanobacteria, weak illuminated due to microalgae bloom;

- *hypertrophic*, greatly excessive nutrient inputs, excessive primary production, the oxygen concentration decreases with depths and the deep water layer is anoxic, slow illuminated, higher turbidity, fish killing possible in summer and under winter ice.

2.2. Climate change

2.2.1. Introduction

Both natural and human factors change the earth's climate. The natural factors which cause the changes in climate are the modifications in the earth's orbit, alterations in the solar activity, or volcanic eruptions. Since the Industrial Revolution began around 1750, human activities have contributed substantially to climate change by adding greenhouse gas emissions including water vapors (H_2O), carbon dioxide (CO_2), methane (CH_4), nitrous oxide (N_2O), and several others which have caused the earth's surface temperature to rise. Atmospheric CO_2 concentrations have increased by more than 40% since preindustrial times, from approximately 280 parts per million by volume (ppmv) in the eighteenth century to 396 ppmv in 2013. **Figure 1** shows the variations of the atmospheric CO_2, CH_4, N_2O per year [7, 16].

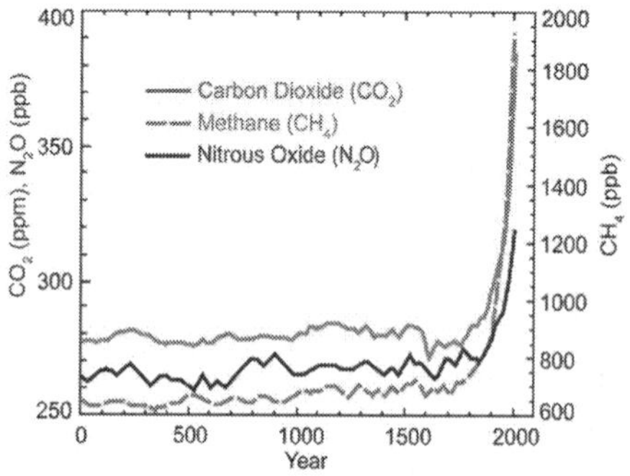

Figure 1. Atmospheric CO_2, CH_4, N_2O variations per year [7].

2.2.2. *Effect of climate change*

Freshwater resources are vulnerable to climate change; warming of the climate system increases global average air and influences the hydrological cycle [17]. Climate change associated with the water cycle (**Figure 2**) includes water body and land temperature increase [18], accelerated glaciers melting, decreasing the surface of water and land occupied with snow, increased evaporation and level of lakes water reduction, increased level of coastal marine and ocean and inundation, wetland loss by sea level rise, changes in the seasonal distribution and amount of precipitation, increasing precipitation intensity sometimes as extreme weather—storms, changes in the balance between snow and rain, increasing nutrients' concentration by soil washing and soil erosion [19], increasing acidity in rivers, lakes, seas, and oceans [20]. **Figure 2** shows the conceptual diagram visualizing the main components of climate change and their major effects on freshwaters [21].

Waters with similar effect filters, however, should respond similar to climate variability. Hydrodynamic patterns are influenced largely by the depth and size of the lake affecting the annual heat budget, temperature stratification during summer and winter, the concentration of oxygen in the hypolimnion, salt solubility, and availability of nutrients. The retention time (a factor depending on morphometry and through-flow) determines if internal or external processes dominate.

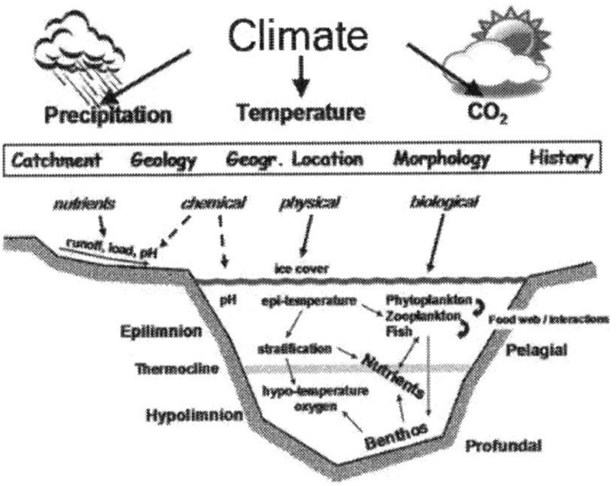

Figure 2. Conceptual diagram visualizing the main components of climate change and their major impacts on freshwaters [21].

Warming of the atmosphere will lead to warmer and wetter winters, and hotter and drier summers. Large quantities of precipitation are expected in winter and spring which will come down as rain rather than snow. The flow regime in stream and rivers changes reaching the maximum flow faster than usual. Larger runoffs combined with more frequent extreme rainfalls may result in floods, increased erosion, and wash out of nutrients, which ultimately lead to the eutrophication of rivers and lakes.

In subtropical and tropical regions, storms and floods occur during rainy seasons. In temperate zones, the summer temperatures increase, the stream and river flow decrease, and the period of thermal stratification extends. In the warmer lakes, the oxygen is less available because the solubility of this element declines with increasing temperature. Decreasing oxygen concentration with increasing temperature and increasing the decomposition rates of organic compounds will increase the consumption of oxygen, which may lead to deoxygenating in deeper parts of lakes. In glaciated regions, the discharge will first increase due to more melt water and later decrease when glaciers have disappeared [19, 22, 23].

3. Stressors' effect on eutrophication characteristics parameters

The stressors' effects can be well highlighted by the mains parameters: temperature, pH, Secchi disc transparency (SD), chlorophyll *a* (CHL), dissolved oxygen (DO), total phosphorus (TP), total nitrogen (TN), and plankton populations [24].

3.1. Temperature and aquatic stratification

Usually, the atmospheric temperature influences the temperature of natural water bodies and these two together depend on the geographical location and meteorological conditions (rainfall, humidity, wind velocity, etc.). The water temperature varies with depths, at the surface the temperature is higher than at greater depths, a phenomenon known as aquatic stratification. The water temperature influences the chemical, physical, and biological processes. Gas solubility decreases with increase in temperature, and biochemical activity doubles every 10°C of temperature increase. Thus, this kind of heating may cause a thermal impact on receiving water bodies and may influence the whole native community [25, 26].

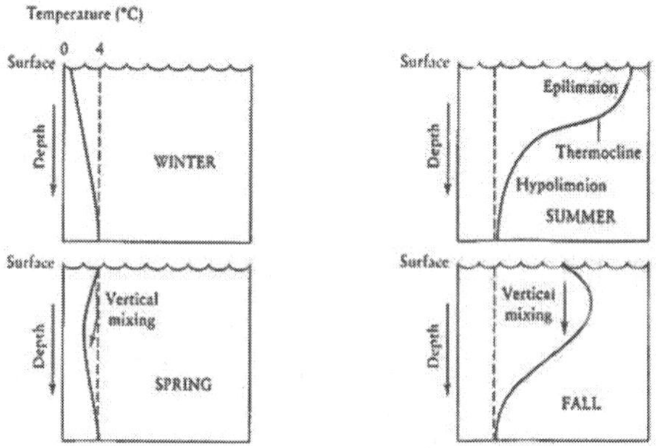

Figure 3. Temperate lakes are thermally stratified in the summer but mix each spring and autumn [28].

Because the lakes and reservoirs are deep enough, they stratify, generating "layers" of water with different physical characteristics. The large differences in density between layers of water determine thermal stratification. The density is influenced by temperature, so at about 3.98 °C the water is most dense (heaviest). The water stratification is seasonal. In the spring time, in temperate climates, immediately after the ice melts, the surface water beings to warm to 0°C. The increasing density of the warming water along with wind cause this surface water to sink and mix with the deeper water, a process called spring turnover. In this period, the water column is cold and has approximately the same temperature (**Figure 3**). During the late spring and summer, the sunlight is absorbed in the water column, heating up with the air average daily temperature increases. In the absence of wind, temperature decreases exponentially with the depth. The lake is now stratified into three layers of water, termed summer stratification. The upper layer, called the epilimnion zone, is in contact with the atmosphere and seasonal climate factors variations, is warm, well-mixed, has a higher pH and higher dissolved oxygen. The thermocline is a plane where the greatest water temperature changes and is very resistant to wind mixing. Hypolimnion starts beneath the metalimnion, extending to the lake bottom, is the coldest layer of a lake in summer, and the warmest layer during winter, usually dark, receives insufficient irradiance (light) for photosynthesis to occur and relatively undisturbed. In deep, temperate lakes, the bottom-most waters of the hypolimnion are typically close to 4°C throughout the year [26, 27].

Epilimnion is cooling down during autumn when decreasing the difference of the density to the hypolimnion. When the temperature of surface and bottom waters and density are approximately the same, winds can mix the entire lake. In winter, the surface water is cooling until it freezes; thus it appears less distinct density stratification because the density difference between 0 and 4°C water is quite small; most of the water column is isothermal at a temperature of 4°C, which is denser than the colder, lighter water just below the ice [26, 29].

Blue-green algae tend to dominate warmer waters while green algae do better under cooler conditions. The toxicity of unionized ammonia is also related to warmer temperatures [30]. The maximum specific growth rate is in the range 5–40°C for members of the Chlorophyta and Bacillariophyta [31]. The optimal temperature for phytoplankton cultures is generally between 20 and 24°C, most commonly cultured species of microalgae tolerate temperatures between 16 and 27°C [32].

3.2. pH

The pH value is a measure of water acidity or alkalinity and the number expresses the concentration of hydrogen ions indirectly and is expressed to the pH scale (measured on a scale of 0–14). Water pH changes are governed by the amount of free CO_2, carbonates, and bicarbonates and are accompanied by the changes in other physicochemical aspects that in turn influence the quality of water. Algal and macrophytes mass increase by the photosynthesis act, the CO_2 increase, in dissolved state, as results of the respiration processes and decomposition of organic matter, reduce the pH [33].

$$CO_2 + H_2O \leftrightarrow H_2CO_3 \tag{1}$$

$$H_2CO_3 \leftrightarrow HCO_3^- + H^+ \tag{2}$$

$$HCO_3^- \leftrightarrow CO_3^{2-} + H^+ \tag{3}$$

The buffering system $CO_2/HCO_3^-/CO_3^{2-}$ maintains pH around the neutral level. Depending on the current pH level operates these equations in both directions. When pH increases Eqs. (2) and (3) shift to the right.

The presence of high alkalinity (>100 mg/l) represents considerable buffering capacity and reduces the effects of both photosynthesis and decay in producing large fluctuations in the pH [30].

A minor change in the pH of water determines increasing solubility of phosphorus and other nutrients—making them more available for plants. Increasing accessible nutrients' quantity, determine increasing demand for dissolved oxygen and creates a eutrophic lake where other organisms living in the water become stressed.

The properly pH range for most fish is between 6.0 and 9.0 with a minimum alkalinity of 20 mg/l, with ideal $CaCO_3$ levels between 75 and 200 mg/l, the pH range for most cultured algal species is between 7 and 9, with the optimum range being 8.2–8.7 [20, 32, 33].

In acidic waters' conditions, only some plants and animals survive. Generally, the younger exemplars of most species are more sensitive to environmental changes. **Figure 4** shows the minimum pH level for different species of fish, shellfish, or the insects which they can tolerate the same amount of acid; for example, frogs can tolerate water that is more acidic (i.e., has a lower pH) than trout [20].

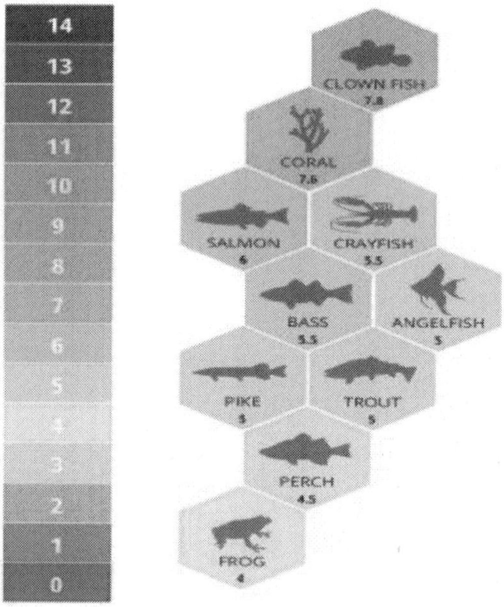

Figure 4. Recommended minimum pH level to survive for different species of aquatic organisms [20, 33].

3.3. Dissolved oxygen

Dissolved oxygen refers to the level of free, noncompound oxygen present in water. Dissolved oxygen concentration is affected by diffusion and aeration, photosynthesis, respiration, and decomposition. The source of the oxygen in the water is the dissolved oxygen from the air and the primary production (photosynthesis process). Depending on the atmospheric conditions, the oxygen enters the water slowly, diffuses quickly by aeration caused by wind, rapid waterfalls, groundwater, etc. The atmospheric temperature and implicit water temperature influence the water oxygen content, and the dissolved oxygen concentrations decrease as temperature increases.

http://dx.doi.org/10.5772/64665

The consumption of oxygen in the lake is the result of two processes: the oxidative and biochemical decomposition processes and respiration (animals, plants, and microbes consume oxygen). The dissolved organic matter is oxidized and the oxygen is taken up by purely chemical oxidation, photochemical oxidation by UV light [25].

Daily and seasonal fluctuations in DO may occur in response to algal and bacterial action. The biological activity increases during the spring, summer, and fall when the photosynthetic activity is high. During the summer and winter, most lakes in the temperate climate are stratified. The combination of thermal stratification and biological activity causes patterns in the water chemistry. The major zones of the lake in relating with oxygen concentration and biological activity are the following [35]:

- The trophogenic zone—the upper stratum of a lake in which photosynthetic production predominates;

- The tropholytic zone—the aphotic deep stratum of a lake where decomposition of organic matter predominates;

- The photic zone—the upper stratum of a lake which receives light input (greater than 1% of surface radiation);

- The aphotic zone—the lower stratum of a lake in which there is no light (less than 1% of surface radiation).

The dissolved concentration varies by season and depth. At turnover (both spring and fall), the O_2 in the water is near 100 % saturation (12–13 mg/l at 4°C at sea level pressure).

During summer stratification, in an oligotrophic lake, the oxygen concentration at depth is influenced by physical processes. The absolute concentration of oxygen decreases in the warmer waters (epilimnion) and increases in the cooler waters (metalimnion and hypolimnion). In a eutrophic lake, the oxygen concentration at depth is influenced by biological processes; the oxidative processes (decomposition and respiration) result in the consumption of oxygen (the dissolved organic matter is oxidized and the oxygen is taken up by purely chemical oxidation, photochemical oxidation by UV light) and the production of oxygen by photosynthesis (primary production):

$$6CO_2 + 6H_2O \rightarrow C_6H_{12}O_6 + 6O_2 \quad (4)$$

Circulation (mixing) and primary production in the epilimnion maintain the oxygen at near 100% saturation. Oxidative consumption reduces the oxygen concentrations in the hypolimnion and the oxygen concentrations have less than 100% saturation [35].

Figure 5 shows the variation of temperature and DO with season and depth in lake water. The amount of dissolved oxygen needed varies from creature to creature. Salmon cannot reproduce at a pH levels below 6 mg/l [36]. Coastal fish avoid areas where the DO is below 3.7 mg/l. Bottom feeders, crabs, oysters, and worms need minimal amounts of oxygen (1–6 mg/l), while shallow water fish need higher levels (4–15 mg/l) [37].

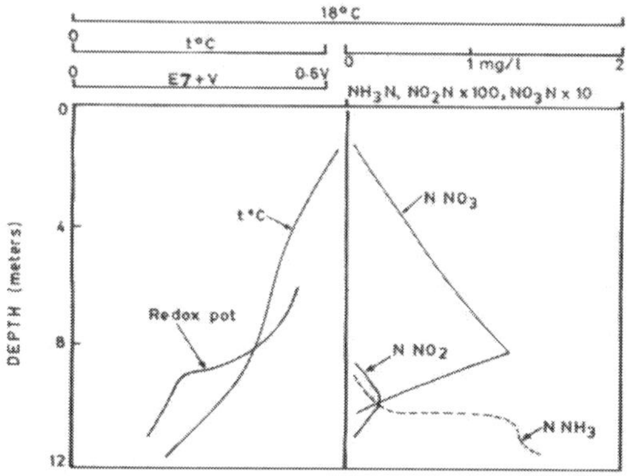

Figure 6. Distribution on depth of the ammonia, nitrate, nitrite, redox potential, and temperature in a temperate lake [43].

3.6. Phosphorus compounds

Phosphorus is one very important macronutrient, which limits the primary production in freshwater. Though phosphorus is the 11th most abundant mineral in the earth's crust, it does not exist in a gaseous state. Phosphorus (P) is an essential nutrient for life, playing a role in the deoxyribonucleic acid (DNA), ribonucleic acid (RNA), adenosine diphosphate (ADP), and adenosine triphosphate (ATP).

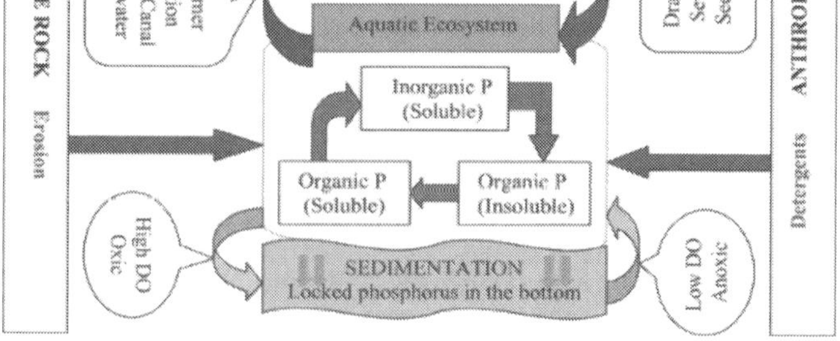

Figure 7. Natural phosphorus cycling between the source, terrestrial, and aquatic ecosystems [6].

http://dx.doi.org/10.5772/64665

The phosphorus in the natural water body is provided by anthropogenic (industrial and agricultural sources) and natural sources. The phosphorus increase is caused by domestic wastewater (detergents and soaps, pesticides, food wastes, and human metabolic waste) [49, 50], food processing industries (meat, vegetable, and cheese processing) [51], distillery, synthetic and natural (cow dung, pig dung, and poultry manure) fertilizers used in agroecosystem [52], agricultural runoff and domestic sewage, phosphate mines [53], and it is very slow, being largely insoluble from mineral matter of rocks (**Figure 7**) [4].

The quantities of phosphorus entering the surface drainage vary with the amount of phosphorus in catchment soils, topography, vegetative cover, quantity and duration of runoff flow, land use, and pollution.

The total phosphorus in aquatic systems, occurs in three forms: inorganic phosphorus (orthophosphate and polyphosphate), particulate organic phosphorus, and dissolved organic phosphorus (soluble and insoluble). The dissolved phase includes inorganic phosphorus, organic phosphorus excreted by organisms, and macromolecular colloidal phosphorus.

Particulate matter includes living and dead plankton, precipitates of phosphorus, phosphorus adsorbed to particulates, and amorphous phosphorus [54]. Total phosphorus concentrations greater than 30μg/l cause algal blooms in lakes and reservoirs.

Aquatic plants require inorganic phosphate, orthophosphate ions (PO_4^{3-}) for nutrition [55]. This form of phosphate is transferred to consumers and decomposes as organic phosphate soluble and insoluble [35].

The deposition of phosphorus into lake sediments occurs by mechanisms such as:

a. sedimentation of phosphorus minerals imported from the drainage basin;

b. adsorption or precipitation of phosphorus with inorganic compounds;

c. uptake of phosphorus from the water column by algal and other attached microbial communities [56].

3.7. Total nitrogen to total phosphorus ratio, TN/TP (redfield ratio)

Nitrogen and phosphorus are two nutrients, which are necessary for microorganism's growth. The nitrogen can be present in three species: nitrate, nitrite, and ammonia, all species are highly soluble in aquatic environment. If nitrogen is in low concentration, the microorganisms can use the nitrogen from atmosphere; also the nitrogen cannot be limited in aquatic systems. The phosphorus is most important nutrient after nitrogen, its concentration controls the plants growth, and it can be easily uptake by precipitation. The total nitrogen to total phosphorus ratio parameter indicates the stage of plant growth [57, 58].

The ratio of nitrogen:phosphorus 10:1 is ideal for aquatic plant growth, the ratio higher than 10:1 indicates phosphorus limited systems; and nitrate accumulates in abundance in water and the ratio less than 10:1, nitrogen limited systems; nitrate will be used soon as input in the water body [30, 38, 59].

3.8. Chlorophyll *a* (CHL)

Chlorophyll *a* is used as a trophic state indicator. It indicates the ratio between planktonic primary production and algal biomass. The algal biomass generates the main problems resulting from eutrophication. It is easier to measure the value of CHL, the algal biomass. Chlorophyll *a* presents a great variability of the cellular chlorophyll content (0.1–9.7% of fresh algal weight), which is influenced by algal types. Seasonally, a great variability in individual cases can be expected due to composition of species, nutrient availability, and light conditions [3, 61].

3.9. Plankton populations

The general effect of eutrophication on the trophic chain consists of excessive growth phytoplankton and macrophytes vegetation, shift to bloom-forming algal species, which might be toxic or inedible, green or brown coloration of the water, increase in the biomass of benthic and epiphytic algae, change in the species composition of macrophytes vegetation, increase of consumer species biomass, increase of fish killing incidence, reduction in species diversity especially macrophytes, frequent occurrence of low dissolved oxygen events (particularly overnight), large pH changes [6, 14].

The major consequence of eutrophication concerns is oxygen availability. By daylight, photosynthesis phytoplankton produces oxygen and biomass and at night, organisms (animal and plants) and microorganisms by respiration and microorganisms by aerobic decomposing (oxidation) of the dead biomass, consume the oxygen. When the all oxygen will be consumed the oxygen from the sulfate will be used by the anoxic bacteria, will release sulfur which will capture the free oxygen still present in the upper layers and in the deep layer there will be accumulated hydrogen sulfite, which has the smell of rotten eggs.

The changes in the water will lead to important changes in the plankton population. Macroalgae, phytoplankton (diatoms, dinoflagellates, chlorophytes), and cyanobacteria (blue/green algae) will experience excessive growth; some of these organisms can release toxins in the water and be toxic themselves. Gelatinous aggregation that floating on the water surface can be produced by blue-green algae and diatoms. In 1982 and 1983, large amounts of gelatinous aggregations were observed on the Aegean Sea [62].

Macroalgal proliferations, the massive developments in spring and summer, also called green tides, are repeatedly observed in the marine environment. The species implicated are frequently from the genus Ulva, Monostroma, Enteromorpha, Elodea Myrisphyllus in fresh waters, Chaetomorpha, and Cladophora. The increasing amount of these type of macroalgae determine the decreasing amount of much more interesting species for biodiversity (autochthonous long-living) such as Fucus. Accumulation of large amounts of these species on beaches can induce numerous nuisances including odor, making it impossible to be used [63].

Most sensitive to oxygen availability, the zooplankton (fish and shellfish, animals with and without limited active locomotion, etc.) may die in oxygen limitation or in water with excessive alkalinity (intense photosynthesis), or toxicity from dangerous metabolic produces (cyanotoxins) or cells themselves of cyanobacteria and other microorganisms.

http://dx.doi.org/10.5772/64665

Humans or animals may be exposed to toxins through the consumption of contaminated drinking water, direct contact with fresh water or the inhalation of aerosols. Toxic compounds can be found free in the water or are cells bound. The normal processes used in treating water for drinking purposes are not efficient so as to remove the free toxins from the water.

Toxins induce damage in animals and humans by acting at the molecular level and consequently affecting cells, tissues, and organs. The main toxin groups include hepatotoxins, neurotoxins, and dermatotoxins, which produce the cyclic peptides, alkaloids, and lipopolysaccharides. The nervous, digestive, respiratory, and cutaneous system may be affected [63].

The symptoms observed on mice using acute doses of hepatotoxins are liver injury and death from liver hemorrhage and cardiac failure within a few hours of exposure. Chronic exposure induces liver injury and promotes the growth of tumors, and cancer. The species of microorganisms that cause the toxic effect are Microcystis, Schizotrix, Plectonema Phormidium, Lyngbia, *Cylindrospermopsis raciborskii,* Anabaena, *Planktothrix agardhii,* Aphanizomenon, Oscillatoria, and Spirula.

Neurotoxins affect the mice and aquatic birds by causing death in a few minutes through respiration arrest. Anabaena, Oscillatoria, Aphanizomenon, Lyngbia are the species responsible for neurotoxins production. Dermatotoxins induce irritant and allergic response in tissues by contact. Lyngbia, Schizothrix, Oscillatoria are most important species which produce the dermatotoxins.

In marine water, over 40 algal species produce the toxins the most important microalgae; Dinophysis, Alexandrium, Gymnodinium, Prorocentrum, Pseudonitzschia (diatoms) are frequently observed and represent a risk for seafood consumers.

The effects include [63]:

- Amnesic shellfish poisoning (ASP), mental confusion and loss of memory, disorientation, and coma;

- Neurotoxic shellfish poisoning (NSP), muscular paralysis, state of shock, and sometimes death;

- Venerupin shellfish poisoning (VSP). Intoxication leads to gastrointestinal, nervous, hemorrhagic, hepatic symptoms and in extreme cases delirium and hepatic coma.

4. The trophic status assessment

4.1. Simple lake characterizations

Trophic state of a lake can be determined by simply observing its basic characteristics (**Table 1**). More profound approaches of trophic state require analysis of key parameters such as phosphorus, nitrogen, chlorophyll *a*, and Secchi depth [65–68]. **Table 1** shows the trophic state classification based on simple lake characterization [69, 70].

Characteristic	Eutrophic state	
	Oligotrophic	Eutrophic
Total aquatic plant production	Low	High
Number of algae species	Many	Few
Characteristic algae groups	Parse	Abundant
Oxygen in hypolimnion	Present	Absent
Characteristic fish	Deep-dwelling cold water fish such as trout, salmon, and cisco	Surface-dwelling, warm water fish such as pike, perch, and bass; also bottom-dwellers such as catfish and carp
Secchi depth	7.5 m	1.5 m

Table 1. Trophic state classification based on simple lake characteristics [69].

4.2. Trophic state per nutrients, primary productivity and Secchi disc parameters

Ecosystems can be described at different trophic states using grow-limiting nutrients, primary productivity and Secchi disc parameters. **Table 2** shows the average characteristics of lakes, streams, and coastal marine waters of different trophic states.

Water body	Tropic state	TN, mg m^{-3}	TP, mg m^{-3}	CHL, mg m^{-3}	SD, m
Lakes	Oligotrophic	<350	<10	<3.5	>4
	Mesotrophic	350–650	10–30	3.5–9	2–4
	Eutrophic	650–1200	30–100	9–25	1–2
	Hypertrophic	>1200	>100	>25	<1
				Suspended CHL, mg m^{-3}	Benthic CHL, mg m^{-3}
Streams	Oligotrophic	<700	<25	<10	<20
	Mesotrophic	700–1500	25–75	10–30	20–70
	Eutrophic	>1200	>75	>30	>70
				Suspended CHL, mg m^{-3}	SD, m
Streams	Oligotrophic	<260	<10	<1	>6
	Mesotrophic	260–350	10–30	1–3	3–6
	Eutrophic	350–400	30–40	3–5	1.5–3
	Hypertrophic	>400	>40	>5	<1.5

Table 2. Average characteristics of lakes [66], streams [67], and coastal marine waters [68] of different trophic states [13].

To increase the efficiency of a lake management program is used a more sophisticate trophic state index to provide more and complete information about the water state. The characterization of trophic status has been conducted using the following: the Carlson's trophic state index, the TSI (Carlson's index).

4.3. Carlson's trophic state index

The trophic state is an absolute scale which describes the biological condition of the water body. The trophic state (TSI) is defined as the total weight of living biological material (*biomass*) in a water body at a specific location and time, a biological response to forcing factors such as nutrient additions [71]. The TSI is the interrelationship between the variables which can be used to identify certain conditions in the lake which are related to the factors limiting the phytoplankton biomass [72]. The effect of nutrients is modified by factors such as season, grazing, mixing depth, etc. For characterizing the trophic state of lakes independent of climate exchange, there were defined the trophic state index (TSI)—Carlson's index Secchi depth, chlorophyll *a*, and total phosphorus; these are three variables which can therefore be used to classify the water body [73]. Three linear regression models are used to calculate the trophic state index and the classified water body: the Secchi disk, TSI(SD); chlorophyll pigments, TSI(CHL); and total phosphorus, TSI(TP). The simplified equation used is presented below [73]:

$$\mathrm{TSI}(\mathrm{SD}) = 60 - 14.41\ln(\mathrm{SD}) \tag{5}$$

$$\mathrm{TSI}(\mathrm{CHL}) = 9.81\ln(\mathrm{CHL}) + 30.6 \tag{6}$$

$$\mathrm{TSI}(\mathrm{TP}) = 14.42\ln(\mathrm{TP}) + 4.15 \tag{7}$$

where TSI(SD) is the trophic state index depending on the Secchi depth, the values of SD is in meters;

TSI(CHL) is the trophic state index depending on the chlorophyll *a* concentration, CHL (μg/l); TSI(TP) is the trophic state index depending on the total phosphorus concentration, TP (μg/l).

Parameter	Oligotrophic	Mesotrophic	Eutrophic	Hypertrophic
Transparency (Secchi depth), SD, m	>4	2–4	2–0.5	0.5–0.25
Total phosphorus, TP, μg/l	<12	12–24	24–96	96–389
Chlorophyll, CHL, μg/l	<2.6	2.6–20	20–56	56–155
TSI	30–40	40–50	50–70	>80

Table 3. Assessment criteria for lake Trophic status (SD, TP, CHL, TSI) [9].

More used is the averaging TSI value, which characterizes the central tendency of the trophic state [73–75]. **Table 3** shows the assessment criteria for the lake trophic status regarding the averaging TSI, the Secchi depth, chlorophyll *a*, and total phosphorus concentration [8].

TSI values	Trophic status	Attributes
<30	Oligotrophic	Clear water, oxygen throughout the year in the hypolimnion
30–40	Oligotrophic	A lake will still exhibit oligotrophy, but some shallower lakes will become anoxic during the summer
40–50	Mesotrophic	Water moderately clear, but increasing probability of anoxia during the summer
50–60	Eutrophic	Decreased transparency, warm-water fisheries only
60–70	Eutrophic	Dominance of blue-green algae, algal scum probable, extensive macrophyte problems
70–80	Hypereutrophic	Heavy algal blooms possible throughout the summer
>80	Hypereutrophic	Algal scum, summer fish killing, few macrophytes

Table 4. Carlson's trophic state index values and classification of lakes [76].

TSI results could be analyzed using Carlson's scale. This is divided into four steps regarding lake productivity: oligotrophic (least productive), mesotrophic (moderately productive); eutrophic (very productive), and hypereutrophic (extremely productive). In natural condition at largely variation of meteorological parameters, a simple interpretation of trophic state of lake water is not enough.

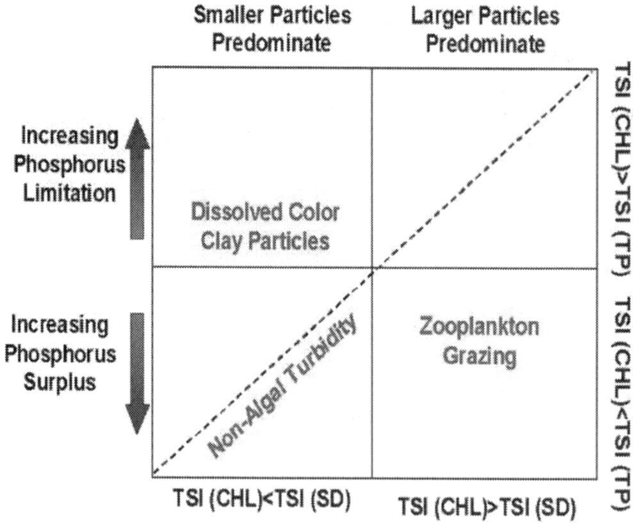

Figure 8. A representation of possible explanations of deviations of the trophic [74].

For complex characterization of natural water must to account of systematic deviations of the simple presentation like in **Table 4**, reported of Carlson in 1992. **Figure 8** illustrates the deviations of TSI(CHL) – TSI(TP) and TSI(CHL) – TSI(SD), and are simultaneously plotted on a single graph, that completes the interpretation of trophic state of natural water. The possibilities are illustrated in **Figure 8** [74]

4.4. Case study: Snagov Lake trophic stage assessment

The Snagov Lake is a natural lake located at 25–30 km North from Bucharest, in Ilfov County, Romania. It is an important natural lagoon on the inferior Ialomita river course with its 5.75 km^2 surface, 16 km length, and 9 m maximum depth, it is included in national patrimony as natural reservation (**Figure 9**).

The lake water sources are the underground waters and in small part snow and rain waters. As consequence, the water level is relatively constant except in winter and autumn [77].

Samples were collected in 2015 during three annual campaigns: April, July, and October. The duplicate of samples were collected from three sampling points than were chosen to monitor the Snagov Lake: input of Lake-Antena Tancabesti, middle of lake—Complex Pacea and output of lake Santu Floresti (**Figure 9**).

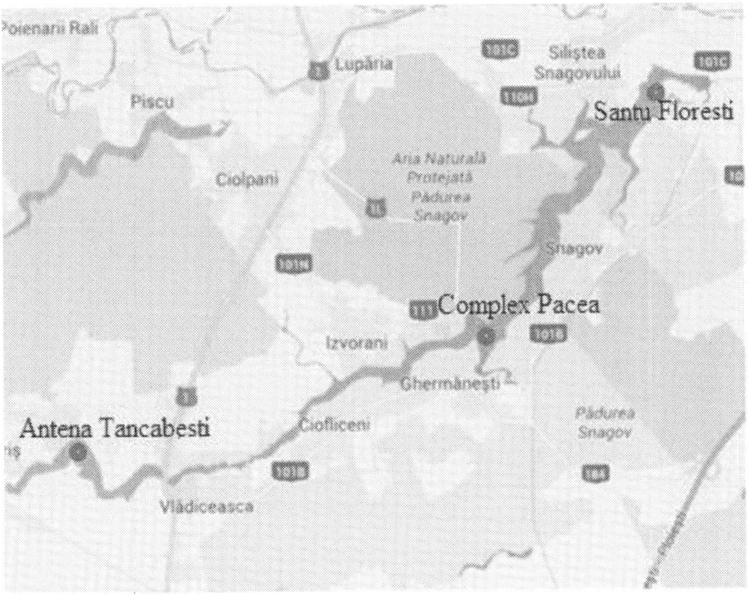

Figure 9. Sampling sites to Snagov Lake: input-Antena Tancabesti, middle-Complex Pacea, and output-Santu Floresti [74].

There were analyzed temperature (*T*), pH, transparency Secchi depth (SD), total nitrogen (TN), total phosphorus (TP), chlorophyll *a* (CHL), dissolved oxygen (DO), turbidity (Tr), total suspended matter (Ts) (**Table 5**).

We calculated the trophic state index (Carlson's index), TSI(SD), TSI(CHL), and TSI(TP) using Eqs. (5)–(7) and there average values TSI and using **Figure 7** and **Table 4** the state of lake was characterized. **Table 6** shows the characterization of Snagov Lake in time in sampling points.

Table 6 shows the evolution of water quality of Snagov Lake in time and in space, at input loaded with nutrient in organic and inorganic matter like smaller particles that involve an excessive development of algae and inorganic matter sedimentation in the middle zone of lake until output when the biological activity slowly decreasing and water quality is slightly improvement, all of this in the eutrophic-hypertrophic state of lake. With this evaluation system can identify the status of the lake and can take necessary measures to improve water quality.

Parameter	Input Antena Tancabest			Middle Complex Pacea			Output Santu Floresti		
	Data			Data			Data		
	April	July	October	April	July	October	April	July	October
T, °C	15	29	16	14	28	17	16	29	16
pH	8.4	8.7	7.9	8.3	7,86	7.74	8,5	8,1	7,6
Ts, mg/l,	21.6	32.2	60.8	21.6	19.6	55.2	23.4	40.2	45.2
SD, m	0.5	0.5	0.5	0.8	0.9	0.9	2	0.45	0.9
Tr, NTU	10	51	32	10.9	5.5	11	18	45	8
DO, mg/l	14.8	21.1	16.8	8.7	10.9	9.4	11.0	9.3	11.5
TN, mg/l	1.7	1.99	0.43	1.47	1.41	0.37	1.85	1.9	0.37
TP, mg/l	0.14	0.23	0.09	0.08	0.14	0.05	0.14	0.08	0.06
TN/TP	12	8.7	4.8	18.8	10.1	7.4	13.2	13.8	6.2
CHL, μg/l	2.4	65.2	58.5	3.6	28.4	22.9	2.37	23.5	11.7
TSI(SD)	70	70	70	63.2	61.5	61.5	50	71.5	61.5
TSI(TP)	75.7	82.5	68.2	67.3	75.4	60.6	75.4	67.3	63.2
TSI(CHL)	39	71.6	69.4	43.2	63.4	61.3	53.8	61.6	54.7
TSI	61.6	74.7	69.2	57.9	66.7	61.1	60	66.8	60

Table 5. Average parameter values for Snagov Lake characterization.

Data	Sampling point	TSI	Relation between TSI variables	Eutrophic stage	Attributes
April	Input of lake	61.6	TSI(TP) > TSI(SD) > TSI(CHL)	Eutrophic	Nonalgal and algal turbidity, smaller particles predominate, temperature lower that optimum temperature of algal growth, close to optimal ratio TN/TP
	Middle of lake	57.9	TSI(TP) > TSI(SD) > TSI(CHL)	Eutrophic	Algal and nonalgal turbidity, lower boundary of classical eutrophy decreased transparency, temperature lower that optimum temperature of algal growth
	Output of lake	60	TSI(TP) > TSI(CHL) > TSI(SD)	Eutrophic	Dominance of blue-green algae, algal scum probable, extensive macrophyte problems, temperature lower that optimum temperature of algal growth
July	Input of lake	74.7	TSI(TP) > TSI(CHL) > TSI(SD)	Hypertrophic	Algal bloom, large particulates dominate, light limited productivity, dense algae, and macrophytes, weak nitrogen deficiency, optimum temperature of algal growth
	Middle of lake	66.7	TSI(TP) > TSI(CHL) > TSI(SD)	Eutrophy	Predominant blue green algae, algae turbidity, larger particles predominant, optimum ratio TN/TP, optimum temperature of algal growth
	Output of lake	66.8	TSI(SD) > TSI(TP) > TSI(CHL)	Eutrophic	Algal and non-algal turbidity, smaller particles predominate, slow phosphorus limitation of algal growth, optimum temperature of algal growth, nitrogen limitation
October	Input of lake	69.2	TSI(SD) = TSI(CHL) > TSI(TP)	Eutrophic-Hipereutrophic	Blue-green algae dominate, light attenuation, zooplankton grazing or toxics limit algal biomass, algae death, large particles predominate, nitrogen limitation
	Middle of lake	61.1	TSI(SD) = TSI(CHL) > TSI(TP)	Eutrophic	Dominance of blue-green algae, algal scum probable, extensive macrophyte problems, nitrogen limitation
	Output of lake	60	TSI(TP) > TSI(SD) > TSI(CHL)	Eutrophic	Algal and nonalgal turbidity, smaller particles predominate, nitrogen limit the algae development Temperature lower than optimum temperature of algal growth

Table 6. Values of trophic state index (TSI) (Carlson's index), the state of lake, and characterization of it to input of lake Antena Tâncăbeşti, middle of Lake Complex Pacea, and output of lake Santu Floresti.

5. Conclusions

Continuous enrichment with nutrients from anthropic sources and enhanced by climate change are important stressors for water bodies. Increasing nutrients concentrations especially nitrogen and phosphorus by anthropogenic sources and extreme weather events, associated with increasing temperature involve important physical, chemical, and biological alterations of water quality. The Carlson's trophic index and its systematic deviations, the TN/TP ratio, and temperature can offer the information regarding trophic state of lake and the characterization of water quality.

Author details

Daniela Simina Stefan[1*] and Mircea Stefan[2]

*Address all correspondence to: simina_stefan_ro@yahoo.com

1 Faculty of Applied Chemistry and Materials Science, University Politehnica of Bucharest, Bucharest, Romania

2 Pharmacy Faculty, University Titu Maiorescu, Bucharest, Romania

References

[1] Dokulii MT, Teubner K. Cyanobacterial dominance in lakes. Hydrobiologia, 2000, 438: 1–12.

[2] Brown C, Canfield D, Bachmann Jr R, Hoyer M. Seasonal patterns of chlorophyll, nutrient concentrations and Secchi disk transparency in Florida lakes. Lake and Reservoir Management, 1998, 14(1): 60–76.

[3] Carpenter SR, Caraco NF, Correll DL, Howarth RW, Sharrpley AN, Smith VH. Nonpoint pollution of surface with phosphorus and nitrogen. Ecological Applications, 1998, 8(3): 559–568.

[4] Khan FA, Ansari AA, Eutrophication: an ecological vision. The Botanical Review, 2005, 71(4): 449–482. http://link.springer.com/article/10.1663/0006-8101%282005%29071[0449:EAEV]2.0.CO%3B2.

[5] Karydis M. Eutrophication assessment of coastal waters based on indicators: a literature review. Global Nest Journal, 2009, 11(4): 373–390.

[6] Ansari AA, Lanza GR, Gill SS, Rast W. Eutrophication: causes, consequences and control. In: Ansari AA, et al. (eds.), Eutrophication: Causes, Consequences and Control,

vol. 143. Dordrecht, Heidelberg, London, New York: Springer Science+Business Media B.V., 2011. ISBN 978-94-007-7813-9, ISBN 978-94-007-7814-6 (eBook), DOI: 10.1007/978-94-007-7814-6.

[7] US EPA. Sources, stressors & responses, Nutrients, 2016. http://www3.epa.gov/caddis/ssr_nut_int.html.

[8] Carlson RE, Simpson JA. Coordinator's Guide to Volunteer Lake Monitoring Methods. Madison, USA: North American Lake Management Society, 1996, p. 96. http://www.secchidipin.org/tsi.htm [cited 1st October 2013].

[9] Vollenweider RA. Coastal marine eutrophication: principles and control. In: Marine Vollenweider RA, Marchetti R, Viviani R (eds.), Marine Coastal Eutrophication, 1992.

[10] Pieterse NM, Bleuten W, Jørgensen SE. Contribution of point sources and diffuse sources to nitrogen and phosphorus loads in lowland river tributaries. Journal of Hydrology, 2003, 271: 213–225.

[11] Carpenter SR. Eutrophication of aquatic ecosystems: bistability and soil phosphorus. PNAS, 2005, 102(29): 10002–10005.

[12] Carlson RE. A trophic state index for lakes. Limnology and Oceanography, 1977, 22(2): 361–369.

[13] Smith VH, Tilman GD, Nekola JC. Eutrophication: impacts of excess nutrient inputs on freshwater, marine, and terrestrial ecosystems. Environmental Pollution, 1999, 100: 179–196. http://cedarcreek.umn.edu/biblio/fulltext/t1758.pdf.

[14] US EPA. Carlson's trophic state index. Aquatic Biodiversity, 2007. http://www.epa.gov/bioindicators/aquatic/carlson.html [accessed 17 February 2008].

[15] Hilton J, O'Hare M, Bowes MJ, Iwan Jones J. How green is my river? A new paradigm of eutrophication in rivers. Science of the Total Environment, 2006, 365: 66–83.

[16] IPCC. Climate change 2013: the physical science basis. In: Stocker TF, Qin D, Plattner G-K, Tignor, M, Allen SK, Boschung J, Nauels A, Xia Y, Bex V, Midgley PM (eds.), Contribution of Working Group I to the Fifth Assessment Report of the Intergovernmental Panel on Climate Change. Cambridge, United Kingdom and New York, NY, USA: Cambridge University Press, 2013. http://www.ipcc.ch/report/ar5/wg1/.

[17] Bates BC, Kundzewicz ZW, Wu S, Palutikof JP (eds.), Climate Change and Water. Technical Paper of the Intergovernmental Panel on Climate Change.Geneva: IPCC Secretariat, 2008, 210 pp. ISBN: 978-92-9169-123-4. http://www.ipcc.ch/.

[18] Hurd BH, Callaway M, Smith J, Kirshen P. Climatic change and US water resources: from modeled watershed impacts to national estimates. Journal of American Water Resources Assessment, 2004, 40: 129–148.

[19] IPCC. Fourth Assessment Report: Climate Change Climate Change Working Group II: Impacts, Adaptation and Vulnerability, Erosion and Sediment Transport, 2007.

[49] Sommaruga R, Cande D, Casal JA. The role of fertilizers and detergents for eutrophication in Uruguay. Fresenius Environment Bulletin, 1995, 4: 111–116.

[50] Berbeiri A, Simona M. Trophic evolution of lake Lugano related to external load reduction: changes in phosphorus and nitrogen as well as oxygen balance and biological parameters. Lake Reservoir Resources Management, 2001, 6: 37–47.

[51] Tusseau-Vuillemin MH. Do food processing industries contribute to the eutrophication of aquatic system? Ecotoxicology Environmental Safety, 2001, 50: 142–143.

[52] Penelope RV, Charles RV. Water Resources and the Quality of Natural Waters. London: Jones and Barbett Publishers, 1992, p. 395–399.

[53] Das BK. Environmental Pollution of Udaisagar Lake and Impact of Phosphate Mine, Udaipur, Rajasthan. Indian Environmental Geology, 1999, 38: 244–248.

[54] Kotoski JE. Phosphorus Mini Fact & Analysis Sheet. Madison, WI: Spring Harbor Environmental Magnet Middle School, 1997.

[55] CCME—Canadian Council of Ministers of the Environment. Canadian water quality guidelines for the protection of aquatic life: phosphorus: Canadian Guidance Framework for the Management of Freshwater Systems. In: Canadian Environmental Quality Guidelines. Winnipeg: Canadian Council of Ministers of the Environment, 2004.

[56] Wenzel RG. Detritus, macrophytes and nutrient cycling in lakes. Memorie dell'lstituto Italiano di Idroln'olog'ia, 1990, 47: 233–249.

[57] Reynolds C. Ecology and Phytoplankton. Cambridge: Cambridge University Press, 2006, p. 47–53.

[58] Smith VH, Bennett SJ. Nitrogen:phosphorus supply ratio and phytoplankton community structure in lakes, Archiv fur Hydrobiologie, 1999, 146: 37–53.

[59] Shen De Z, Niu J, Wang X, Wang H, Zhao X. Distribution and transformation of nutrients in large-scale lakes and reservoirs. Springer Science & Business Media, 2013.

[60] Guildford S, Hecky RE. Total nitrogen, total phosphorus, and nutrient limitation in lakes and oceans: is there a common relationship. Limnology and Oceanography, 2000, 45: 1213–1223.

[61] Boyer JN, Kelble CR, Ortner PB, Rudnick DT. Phytoplankton bloom status: chlorophyll-*a* biomass as an indicator of water quality condition in the southern estuaries of Florida, USA. Ecologic Indicators, 2009, 56–67.

[62] Ignatisdes L, Gostsis-Skretas O. A review on toxic and harmful algae in Greek coastal waters (E. Mediterranean Sea). Toxins (Basel), 2010, 2(5): 1019–1037. http://www.mdpi.com/2072-6651/2/5/1019.

[63] Chorus I, Bartram J (eds.), Toxic cyanobacteria in water, London, E & FN Spon. In: Vollenweider RA, Marchetti R, Viviani R (eds.), Coastal Eutrophication. London: Elsevier, 1999, p. 1–20.

[64] Kein G, Perera P. Eutrophication and health. World Health Organization Regional Office for Europe, 2002. http://ec.europa.eu/environment/water/water-nitrates/pdf/eutrophication.pdf.

[65] Vollenweider RA, Kerekes JJ. Background and summary results of the OECD cooperative program on eutrophication. In: Proceedings of an International Symposium on Inland Waters and Lake Restoration. U.S. Environmental Protection Agency, EPA, 1980, p. 440/5-81-010.

[66] Nürnberg GK. Trophic state of clear and colored, soft- and hard-water lakes with special consideration of nutrients, anoxia, phytoplankton and fish. Lakes and Reservoir Management, 1996, 12: 432–447.

[67] Dodds WK, Jones JR, Welch EB. Suggested classification of stream trophic state: distributions of temperate stream types by chlorophyll, total nitrogen, and phosphorus. Water Research, 1998, 32: 1455–1462.

[68] Hakanson L. A review of effect [dose-sensitivity models for aquatic ecosystems]. Internationale Revue der Gesamten Hydro- Biologie, 1994, 79: 621–667.

[69] Rast W, Lee GF. Summary Analysis of the North American (US Portion) OECD Eutrophication Project: Nutrient Loading-Lake Response Relationship and Trophic State Indices. US EPA. Corvallis, OR: Corvallis Environmental Research Laboratory, 1987, p. EPA-600/3-78-008.

[70] Brown T, Simpson J. Urban lake management. Chapter: Determining the Trophic State of Your Lake, 2016.

[71] Naumann E. The scope of chief problems of regional limnology. Internationale Revue der gesamten Hydrobiology, 1929, 21: 423–444.

[72] Carlson RE. More complication in the chlorophyll-*a*, Secchi's disc relationship. Limnology and Oceanography, 1980, 25: 378–382.

[73] Carlson RE. Discussion on "Using differences among Carlson's trophic state index values in regional water quality assessment" by Richard A. Osgood. Water Resources Bulletin, 1983, 19: 307–309.

[74] Carlson RE. Expanding the trophic state concept to identify non-nutrient limited lakes and reservoirs. In: Proceedings of a National Conference on Enhancing the States' Lake Management Programs. Chicago: Monitoring and Lake Impact Assessment, 1992, p. 59–71.

[75] Osgood R. Using differences among Carlson's trophic state index values in regional water quality assessment. Water Resources Bulletin, 1983, 18: 67–74.

[76] Prasad AGD, Siddaraju. Carlson's trophic state index for the assessment of trophic status of two Lakes in Mandya district. Advances in Applied Science Research, 2012, 3(5): 2992–2996. Available online at www.pelagiaresearchlibrary.com.

[77] Snagov lake map. Google map. https://www.google.ro/maps/@44.6850248,26.133245,12z.

[78] REPA Bucharest-Ilfov report, 2010. http://arpmbuc.anpm.ro/Mediu/raport_priv-ind_starea_mediului_in_romania-15.

Chapter 3

Determination of the Effect of Drought Stress on the Seed Germination in Some Plant Species

Nurcan Yigit, Hakan Sevik, Mehmet Cetin and Nur Kaya

Additional information is available at the end of the chapter

http://dx.doi.org/10.5772/63197

Abstract

Especially the use of drought-resistant plant species reduces maintenance and irrigation costs, and plants increase the retention and success to continue its life in arid landscape. In this study, some plant species used have been studied to determine their tolerance to drought stress in gardens and parks in Kastamonu. For this purpose, germination trials have been in conducted -2, -4, -6, and -8 Bar water stress. Landscaping applications commonly used some species such as *Cupressus sempervirens* L., *Ailanthus altissima* (Mill.) Swingle, *Pyracantha coccinea* Roem, *Thuja orientalis*, *Pinus sylvestris* L., *Sophora japonica*, *Cedrus libani* A. Rich., *Acer pseudoplatanus* L., *Pinus brutia* Ten., and *Pinus nigra* Arnold. ssp. *pallasiana* (Lamb.) Holmboe. Their seeds were evaluated different levels of water stress in the germination percentage. PEG 6000 solution was used in the formulation of water stress. The seeds were exposed to constant temperature of 25°C for a period of 35 days at germination cabinet. As a result, this experiment calculated germination in different water stress levels what percentage has fallen, so the least affected by increased water stress was studied to determine the species. Also results showed increased water stress and reduce the percentage of germination in all species. The highest level of water stress -8 Bar, which was also obtained stress level proportional germination values *Pinus nigra* Arnold. ssp. *pallasiana* (Lamb.) Holmboe (64.8%) and *Pinus brutia* Ten. (46.5%).

Keywords: polyethylene glycol, germination percentage, PEG, drought stress, plants, seed

INTECH
open science | open minds

© 2016 The Author(s). Licensee InTech. This chapter is distributed under the terms of the Creative Commons Attribution License (http://creativecommons.org/licenses/by/3.0), which permits unrestricted use, distribution, and reproduction in any medium, provided the original work is properly cited.

1. Introduction

Rapidly increasing population in the world, limitless industrialization process, poor urbanization activities, regional wars, pesticides which are used to increase crops, and unconscious use of fertilization and chemicals such as detergents have started to contaminate the environment, which results in damage for living beings as a result of extensive air, water, and soil pollutions. The use of fossil fuels has been on the increase since the industrial revolution. With the addition of rapid deforestation, these factors have yielded a serious situation nearly beyond prevention [1, 2]. Throughout history, the amount of CO_2 concentration in the air did not exceed 320 ppm. However, current concentration is above 385 ppm and it keeps on increasing [3]. This situation has led to a concern and long debates in relation to the effects of global warming [1].

It is inevitable that climate change manifests its effects all around the world due to global warming. The increase in temperature and changing precipitation are expected to increase water problems, which are already felt in certain regions. It is estimated that there will be changes in the frequency and severity of droughts and floods, which may lead to serious loss of lives and properties throughout Europe [4].

Drought is a phenomenon resulting from certain variables such as precipitation, temperature, humidity, evaporation, and transpiration. Basically, drought refers to water deficit resulting from below-average emergence of natural water assets, which are used by various systems, in some regions for a particular time period [5]. When drought is mentioned, precipitation and water deficit come to one's mind first. It can be defined as having higher levels of water loss due to evaporation than the water supplied by precipitation in a certain region [6].

Twenty-eight percentage of usable territories on the world are affected by drought [5]. Due to greenhouse gas accumulation in the atmosphere, a climate change will take place in the upcoming years, which is likely to leave us with arid and sub-arid territories. These, in turn, will add to the water problems in urban areas, and there will be an increase in the demand for water for both agricultural and drinking purposes. Hence, in addition to the expansion of arid and sub-arid territories, there will be an increase in summer drought period and severity accompanied by desertification, salinization, and erosion processes [7]. Undoubtedly, one of the most vulnerable territories to drought is green and outdoor spaces in cities.

The drought increasing day by day and the thirstiness as a result of this make their destructive effects felt in green fields as it is in all parts of our lives. Almost all of the green fields constituted with the approach of classical landscaping design, which requires great amounts of water especially in our metropolitan cities, were damaged greatly in a few months in which water usage was restricted [8–11].

While the purpose was healing the environment-ambience quality in the applications of landscaping architecture and repairing the damaged environment conditions in previous years, the wise usage of water for the worries depending on the climate change and herbal applications resistant to drought have come to the fore recently [12–16]. The plants used in landscaping fields are desired to be resistant to drought, and this becomes the most important criterion, which affects the choice of plants in some regions even [17–21]. It is of great importance for the wise usage of water that species and origins resistant to drought are identified first and these kinds of species and origins is used for the landscaping designs.

Different methods are used for identifying the resistances of species against drought. One of these methods is to conduct PEG applications on the seeds in different concentrations. PEG applications have been used in many species for identifying their resistance to drought, and it has given very successful results [22, 23].

In this study, it was aimed at identifying the water stress tolerance for some plants, which are used in landscaping works. Trials have been conducted on nine species chosen for this purpose, and the water stress reactions of the species in different levels were determined.

2. Material and methods

In the study, a total of 10 species have been used which are *Cupressus sempervirens* L., *Ailanthus altissima* (Mill.) Swingle, *Pyracantha coccinea* Roem, *Thuja orientalis*, *Pinus sylvestris* L., *Sophora japonica*, *Cedrus libani* A. Rich., *Acer pseudoplatanus* L., *Pinus brutia* Ten., and *Pinus nigra* Arnold. ssp. *pallasiana* (Lamb.) Holmboe and which are frequently used in landscaping applications. The fruits and cones of above-mentioned species were gathered in October in the parks and gardens in Kastamonu, and seeds were attained after cleaning the cones and fruit fleshes. The seeds are except the ones of *Pinus nigra* and *Pinus brutia* which have been subject to stratification for a period of 8 weeks, and then, germination trials have been conducted in -2, -4, -6, and -8 Bar water stress, and in addition to these, control group has been used. The water potential of the germination substrates (0, -2, -4, -6 and -8 Bar) was determined using PEG-6000 solution and prepared as described by Michel and Kaufman [24, 25].

Germination tests were performed in 11-cm-diameter glass petri dishes on two layers of filter paper saturated with water solutions. Fifty pieces of seeds have been placed in each petri dish in a way that the seeds will not have a touch with each other, and the study has been conducted with 4 repetitions. In this way, a total of 9000 seeds consisting of 50 pieces of seeds, 4 repetitions, 5 applications, and 9 species have been used in the study.

The seeds have been subject to germination with the period of 35 days in 25 ± 1.0°C constant–temperature in germination cabinet, and the filter papers have been renewed in each 3 days.

At the end of 35-day period, non-germinated seeds were cut and checked. Unfilled seeds were excluded from the evaluation, and germination percentage (GP) was calculated by proportioning the number of germinated seeds to healthy ones.

Germination percentage results were subjected to factorial variance analysis. The differences between species and the degrees of significance of such differences were revealed. In order to eliminate the misleading effects of the germination capability differences between the species included in the experiment in checking operations (0 Bar) on the analysis results, the values of these operations were proportioned to 100, and cumulative germination percentages (CGP) were calculated. Thus, the differences between the species were tried to be demonstrated more rationally. The data to be subjected to variance analysis were entered as they were proportioned to 100.

SPSS 17.0 statistics program was used to evaluate the data obtained through experiments. When statistically significant differences ($P < 0.05$) were found through analyses of variance, Duncan's test was performed to form homogeneous groups. Duncan's test showed that the operations were in the same or different categories in terms of the measured characteristics [26]. Multivariate analysis of variance (ANOVA) employs two or more factors and attempts to reveal significant differences between the mean scores of many groups according to these factors. In other words, multivariate analysis of variance is used to test the difference between the mean scores of k-dependent groups [27]. In order to interpret study results with more ease, Excel program was used to create graphs.

3. Findings

Table 1 shows the end-of-study values regarding the change in the germination percentages of species under water stress conditions.

Species	Water stress conditions				
	Control	-2 Bar	-4 Bar	-6 Bar	-8 Bar
Pinus brutia	45,2	40,0	36,3	24,5	21,0
Pinus nigra	81,9	71,0	65,5	62,3	53,1
Cupressus sempervirens	62,6	54,3	45,8	36,7	19,7
Ailanthus altissima	60,2	50,0	12,2	1,2	0,0
Pyracantha coccinea	51,2	32,5	26,6	10,2	5,6
Thuja orientalis	70,2	65,3	48,0	35,2	15,6
Pinus sylvestris	72,6	66,7	41,3	28,6	12,3
Sophora japonica	55,5	45,0	30,5	7,1	1,2
Cedrus libani	32,3	30,7	18,0	7,2	2,4
Acer pseudoplatanus	38,7	28,6	20,2	5,1	0,0

Table 1. The change of germination percentages under various water stress conditions.

Considering the values in **Table 1**, the highest germination was observed in *Pinus nigra* with a rate of 81.9% in the control group. It was followed by *Pinus sylvestris* with a rate of 72.6% and *Thuja orientalis* with a rate of 70.2%, respectively. *Cupressus sempervirens* and *Ailanthus altissima* yielded germination higher than 60% in the control group. *Cupressus sempervirens'* percentage was 62.6%, while the rate of *Ailanthus altissima* was 60.2%. Germination percentage of *Sophora japonica* was 55.5%, while the rate of *Pyracantha coccinea* was 51.2%. Germination percentages of *Pinus brutia, Acer pseudoplatanus,* and *Cedrus libani* remained under 50% in the control groups. *Pinus brutia* yielded a germination percentage of 45.2%, and *Acer pseudoplatanus'* rate was 38.7%. The lowest rate belonged to *Cedrus libani* with 32.3%.

Pinus nigra, having the highest germination rate in the control group with a rate of 81.9%, yielded a lower rate, 71%, under -2 Bar water stress level. The rate reduced to 65.5% under -4 Bar, 62.3% under -6 Bar, and 53.1% under -8 Bar. The values in the table show that the highest germination rates belonged to *Pinus nigra* at all experiment levels.

Pinus sylvestris had the highest germination rate following *Pinus nigra* in the control group with a rate of 72.6%. This rate reduced to 66.7% under -2 Bar, 41.3% under -4 Bar, 28.6% under -6 Bar, and 12.3% under -8 Bar. Though *Pinus sylvestris* seeds had the second highest germination rate in the control group, it rapidly reduced depending on water stress levels. It had the second highest value under -2 Bar water stress level; however, it ranked the 4th under -4 and -6 Bar water stress levels, while it became the fifth under -8 Bar water stress level. It is possible to say that *Pinus sylvestris* is more vulnerable than other seeds under examination against the increasing water stress conditions.

Thuja orientalis seeds had a percentage of 70.2% germination in the control group; however, this rate reduced in parallel with the increasing water stress and became 65.3% under -2 Bar, 48% under -4 Bar, 35.2% under -6 Bar, and 15.6% under -8 Bar.

Cupressus sempervirens also experienced a decrease in germination percentage due to increasing water stress. Its germination rate was 62.6% in the control group, but it reduced to 54.3% under -2 Bar, 45.8% under -4 Bar, 36.7% under -6 Bar, and 19.7% under -8 Bar, which was the highest water stress level. *Cupressus sempervirens* ranked 4th in terms of germination percentage in the control group, but it became the 3rd under -8 Bar.

Ailanthus altissima ranked 5th in the control group in terms of germination percentage. However, *Ailanthus altissima* experienced a rapid fall in germination percentage particularly after -4 Bar water stress level. The germination percentage which was 50% under -2 Bar reduced to 12.2% under -4 Bar and displayed 1.2% of germination percentage under -6 Bar. The seed did not germinate under -8 Bar. Accordingly, *Ailanthus altissima* had the lowest germination percentage under -4 and -6 Bar water stress levels.

Sophora japonica showed a germination percentage of 55.5% in the control group. This rate reduced to 45% under -2 Bar, 30.5% under -4 Bar, 7.1% under -6 Bar, and 1.2% under -8 Bar water stress levels.

Another species experiencing a rapid fall in germination percentage due to increasing water stress level was *Pyracantha coccinea*. Its germination percentage in the control group was 51.2%. It first reduced to 32.5% and then to 26.6% due to increasing water stress and reached 10.2% under -6 Bar and 5.6% under -8 Bar.

Though its germination percentage was below 50% in the control group, *Pinus brutia* was one of the species experiencing one of the lowest fall in germination percentage due to increase in water stress. Germination percentage of *Pinus brutia* in the control group was 45.2%, while this rate reduced to 40% under -2 Bar, 36.3% under -4 Bar, and 24.5% under -6 Bar. Having one of the three lowest germination percentages in the control group, *Pinus brutia* had a germination rate of 21% under -8 Bar water stress level, which was the second highest value.

Acer pseudoplatanus was one the species with lower than 50% germination percentage in the control group. Its germination percentage rapidly fell due to increasing water stress. While its percentage in the control group was 38.7%, it reduced to 28.6% under -2 Bar, 20.2% under -4 Bar, and 5.1% under -6 Bar. *Acer pseudoplatanus* was one of the two species which did not germinate under -8 Bar water stress level.

Cedrus libani had the lowest germination percentage in the control group (32.3%). While its germination percentage was 30.7% under -2 Bar, this value reduced to 18% under -4 Bar, 7.2% under -6 Bar, and 2.4% under -8 Bar. According to these rates, *Cedrus libani* had the lowest germination percentage in the control group; however, its germination percentage under -8 Bar was the 4th lowest germination percentage. The **Figure 1** shows the change in the germination percentages observed in the species due to increasing water stress levels.

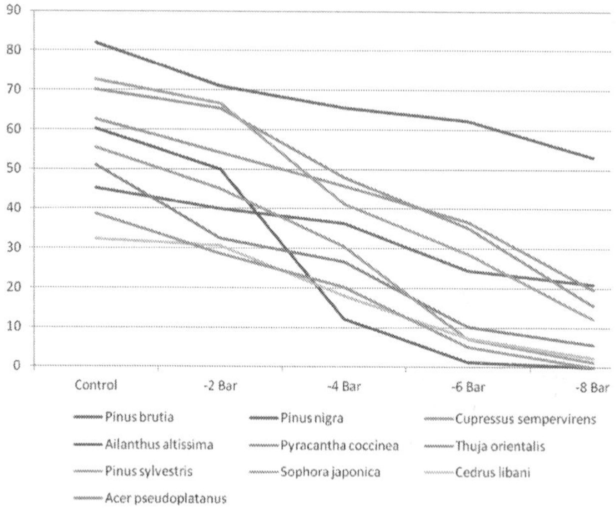

Figure 1. The changes in the germination percentages of the species due to increasing water stress levels.

The purpose of this study was to reveal the changes in the germination percentages of species depending on the water stress and to determine which species are least influenced by increasing water stress. However, not only the germination percentages of species but also to what extent they experience a fall in germination depending on the increasing water stress are of importance in order to make an evaluation in this matter. As a matter of fact, the germination percentage of a species may be low due to its biology or environmental conditions. For instance, Khera and Singh [28] studied various origins in their studies. They reported that *Dalbergia sissoo*'s germination percentage varies between 93 and 21% in different origins. Therefore, there is a need to reveal to what extent germination percentage reduces depending on increasing water stress, which is expected to provide a sounder evaluation of tolerance towards water stress among species and facilitate making comparison between them. Hence, the species in the control group were accepted to have a germination percentage of 100%. The rates of reduction depending on the increasing water stress, in other words, cumulative percentages (CGP) are given in **Table 2**.

Species	Water stress conditions				
	Control	-2 Bar	-4 Bar	-6 Bar	-8 Bar
Pinus brutia	100	88,5	80,3	54,2	46,5
Pinus nigra	100	86,7	80	76,1	64,8
Cupressus sempervirens	100	86,7	73,2	58,6	31,5
Ailanthus altissima	100	83,1	20,3	2	0
Pyracantha coccinea	100	63,5	52	19,9	10,9
Thuja orientalis	100	93,0	68,4	50,1	22,2
Pinus sylvestris	100	91,9	56,9	39,4	16,9
Sophora japonica	100	81,1	55	12,8	2,2
Cedrus libani	100	95,0	55,7	22,3	7,4
Acer pseudoplatanus	100	73,9	52,2	13,2	0

Table 2. The change of the germination percentage compared with the control group.

The values in the table show that the germination percentage reduced in all of the species due to increasing water stress. The minimum change took place in *Cedrus libani, Thuja orientalis,* and *Pinus sylvestris* under -2 Bar water stress level. Germination percentage under -2 Bar water stress level in *Cedrus libani* corresponded to 95% of the one in the control group, while germination percentage rate corresponded to 93 and 91.9% of the values in the control group for *Thuja orientalis* and *Pinus sylvestris,* respectively, under the same conditions. The changes in the germination percentage under -2 Bar water stress level compared with the control group are as follows: *Pinus brutia* 88.5%, *Cupressus sempervirens* and *Pinus nigra* 86.7%, *Ailanthus altissima* 83.1%, and *Sophora japonica* 81.1%. For these species, germination percentage under -2 Bar water stress level was higher than 80% of the control group. This rate was 73.9% for *Acer pseudoplatanus* and 63.5% for *Pyracantha coccinea.*

However, most of the species experienced great falls in their germination percentages starting from -4 Bar water stress level. The minimum change under -2 Bar water stress level was observed in *Cedrus libani, Thuja orientalis,* and *Pinus sylvestris*. Germination percentage of *Cedrus libani* fell to 55.7% of the control group under -4 Bar water stress level. These rates were 68.4 and 56.9% for *Thuja orientalis* and *Pinus sylvestris,* respectively. The species which were least affected by water stress at this level were *Pinus brutia, Pinus nigra,* and *Cupressus sempervirens*. The changes in the germination percentages of the species under -4 Bar water stress level compared with the control group percentages were 80.3% for *Pinus brutia,* 80% for *Pinus nigra,* and 73.2% for *Cupressus sempervirens*. The species which experienced the highest change under -4 Bar water stress compared with the control group are as follows: *Sophora japonica* with

55%, *Acer pseudoplatanus* with 52.2%, Pyracantha *coccinea* with 52%, and *Ailanthus altissima* with 20.3%.

Not a big difference occurred in these results under -6 Bar water stress level. The species had similar ranks to the ranks under -4 Bar water stress level. Once again, the highest values in the comparison of germination percentages to the control group percentages were observed in *Pinus nigra* (76.1%), *Cupressus sempervirens* (58.6%), *Pinus brutia* (54.2%), and *Thuja orientalis* (50.1%). The rate was below 50% in all of the other species. Under -6 Bar water stress level, the rate was 39.4% for *Pinus sylvestris* and 22.3% for *Cedrus libani,* while other species went below 20%. *Pyracantha coccinea* had a rate of 19.9%; *Acer pseudoplatanus* had a rate of 13.2%; and *Sophora japonica* had a rate of 12.8%. The highest proportional fall under -6 Bar water stress level was observed in *Ailanthus altissima,* and germination percentage could only reach 2% compared with the control group.

Under -8 Bar water stress, which is the highest water stress level, *Ailanthus altissima* and *Acer pseudoplatanus* did not germinate. Under these conditions, the changes in germination percentages compared with the control group are as follows 2.2% for *Sophora japonica,* 7.4% for *Cedrus libani,* 10.9% for *Pyracantha coccinea,* 16.9% for *Pinus sylvestris,* 22.2% for *Thuja orientalis,* and 31.5% for *Cupressus sempervirens.* Under the highest water stress level, which is -8 Bar, the highest germination percentages were observed in *Pinus nigra* (64.8%) and *Pinus brutia* (46.5%). The graph showing the reduction rates in the germination percentages of species due to increasing water stress is given in **Figure 2**.

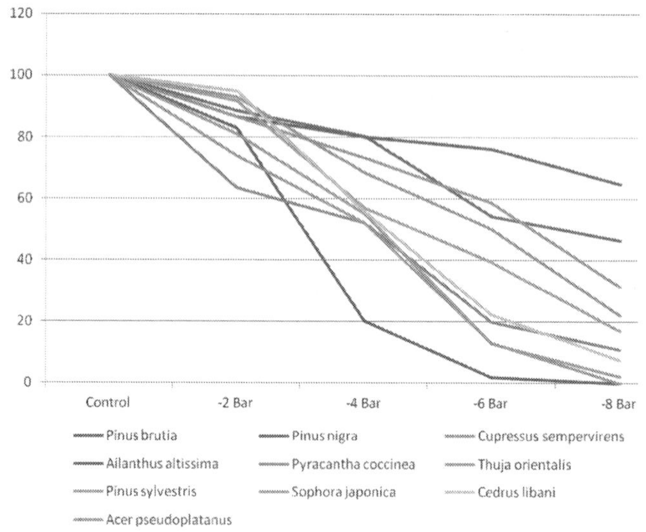

Figure 2. The reduction rates in the germination percentages of the species due to increasing water stress.

4. Discussion and conclusion

In today's modern life, it has been accepted that the presence of plants in cities is an indicator of their quality and inhabitability [29]. Plants reduce the air pollution and noise in their surrounding areas [30–35]. They also increase aesthetic value [36], have a good influence on psychology [37, 38], save energy [39, 40], prevent erosion [41], and decrease the speed of winds. Since they penetrate into soil with their roots, they prevent transportation of soil by precipitation and streams. They also protect wild life and hunting resources. Green and open areas surrounded by plants are important activity areas for both adults and children [42, 43]. In addition, indoor plants increase the productivity of people working in these places [44]. They relieve people psychologically and reduce stress and negative feelings [45–47].

Due to these functions of plants, a lot of issues such as plants' spread areas, [48–52], protection [53–59], production [60–64], tolerance against stress factors [65, 66], use in various areas [67–69], genetic variability [70–72], relationships with environment and other living beings [73–79], and raising awareness about them, as well as their legal aspects [80–82] have become main study areas. Therefore, a lot of studies have been conducted on these subjects.

In addition to these functions of plants, their contributions to the aesthetic aspect of the places they are in should be dwelt on under a separate title. Landscape practices for which various species and varieties are used have gained a distinct importance in the modern world. The desire to use diverse species has led to an intense use of plants outside their natural spread areas. The species which are not part of the natural flora of the region draw more attention when they are used for landscape practices and increase landscape quality. However, these practices also cause such plants to deal with ecological and climatic conditions which they are not used to. Therefore, maintenance and watering costs of such species are higher. However, global warming makes it necessary to have a reasonable and thrifty attitude in the use of water.

It is inevitable that climate change manifests its effects all around the world due to global warming. It is expected that the increase in temperature and changing precipitation will increase water problems. It is expected that there will be changes in the frequency and severity of droughts and floods throughout Europe, which may result in loss of lives and property [83]. Therefore, there is a need to determine species which are tolerant to water stress and use such species in both landscaping and forestation practices.

This study is an attempt to reveal the tolerances of certain plant species, which are commonly used for landscape practices, against drought stress. The results of the study indicate that increasing water stress reduced the germination percentage in all the species examined. Many previous studies have reported similar results for many other species so far. Sevik and Cetin [84] conducted a study to determine species' tolerances towards water stress and reported that the species which were most affected by water stress were *Sophora japonica, Ailanthus altissima,* and *Cupressus arizonica,* whereas the species which were most tolerant against water stress were *Pinus nigra, Cupressus sempervirens,* and *Pinus brutia.*

Falusi et al. [85] analyzed the influence of increasing water stress on the germination percentages of four origins of *Pinus halepensis* and revealed that there were great differences between germination percentages of the origins. The origin, which was least affected by water stress, had a germination percentage of 94.10% in the control group, while this rate was 63.64% under -8 Bar water stress level. The origin, which was most affected by water stress, had a germination percentage of 90.1% in the control group, while this rate reduced to 11.8% under -8 Bar water stress level.

Tilki and Dirik [86] performed experiments on various origins of *Pinus brutia*. They reported that Silifke-origin seeds had a germination percentage of 78.7% in the control group. This rate reduced to 44.8% under -0.4 MPa level. As for the Cehennemdere-origin seeds, they showed a germination percentage of 33.2% in the control group while this rate reduced to 4% under -0.4 MPa level. In this study, germination percentage of *Pinus brutia* was 45.2% in the control group while this rate was 40% under -2 Bar, 36.3% under -4 Bar, and 24.5% under -6 Bar water stress levels.

Boydak et al. [87] conducted a study with *Pinus brutia* seeds of 6 different origins. The germination percentage was observed to be 84.3% in the control group. This rate changed to be 88.7% under -2 Bar. It dramatically reduced after -4 Bar water stress level. Under -4 Bar water stress level, the percentage was 80.6%, whereas it was 55.5% under -6 Bar water stress level. The germination percentage was 25.2% under -5 Bar water stress level and was 29.8% under -8 Bar water stress level, according to what they report.

Ahmadloo et al. [88] conducted a study on *Cupressus sempervirens* and *Cupressus arizonica* and created -2, -4, -6 and -8 Bar water stress levels. They analyzed germination percentages. It was seen that *Cupressus arizonica* yielded 18.75% germination percentage in the control group. This rate reduced to 14.5% under -2 Bar water stress level. It was found to be 10.5% under -4 Bar, 9.25% under -6 Bar, and 7% under -8 Bar water stress levels. Germination percentage of *Cupressus sempervirens* in the control group was 27.75%. Under -2 Bar water stress level, this rate reduced to 18.5%, 18% under -4 Bar, 11.75% under -6 Bar, and 7.5% under -8 Bar water stress levels. These results generally support the results of this study. In this study, germination percentage of *Cupressus sempervirens* L. was 62.6% in the control group. This rate reduced to 54.3% under -2 Bar, 45.8% under -4 Bar, 36.7% under -6 Bar, and 19.7% under -8 Bar, which is the highest water stress level. Therefore, it was seen that germination percentage reduced to 31.5% of the control group under -8 Bar water stress level. Ahmadloo et al. [88] reported this number as 27%.

-5, -10, -15, and -20 Bar water stress levels were created in order to reveal the influence of water stress on the germination percentages of *Acacia catechu, Acacia nilotica, Albizia lebbek, Dalbergia sissoo*, and *Tectona grandis*, and the changes in the germination percentages were observed. At the end of the study, it was seen that the germination percentage which was 90% for *Acacia nilotica* in the control group became 90% under -20 Bar water stress level. Other species experienced a significant fall in germination percentages. *Acacia catechu* had a germination percentage of 94% in the control group, while this rate reduced to 60% under -20 Bar water stress level. *Albizia lebbek* had a germination rate of 59% in the control group, while this rate reduced to 35% under -20 Bar water stress level. The germination percentage for *Dalbergia sissoo* was 93% in the control group, while this rate reduced to 53% under -20 Bar water stress level. The germination percentage of *Tectona grandis* was 22% in the control group, while this rate reduced to 6% under -15 Bar and 0% under -20 Bar water stress levels [28].

Boydak et al. [87] conducted a study with 6 different origins of *Pinus brutia* seeds. The average germination percentage of the seeds was 84.3% in the control group. The rate reduced to 25.2% under -8 Bar water stress level. That study also revealed that origins differed in their tolerances towards water stress. For instance, germination percentage in the 2nd origin was 91% in the control group, and this rate reduced to 40.5% under -8 Bar. In addition, the 5th origin had a germination percentage of 94.5% in the control group, and this rate reduced to 15% under -8 Bar water stress. Proportional values make the results clearer. The 2nd origin's proportional germination percentage was found to be 44.5% under -8 Bar water stress level, while this rate was found to be 15.8% for the 5th origin under the same conditions.

Kaufmann and Eckard [89] stated that water stress at a level of -8 Bar may reduce the germination percentages of Pinus *contorta* and *Picea engelmannii* seeds at a rate of 50%. Djavanshir and Reid [90] evaluated germination percentages in *Pinus ponderosa* and *Pinus elderica.* They concluded that increasing water stress affected the germination percentage; and germination percentage reduced nearly to zero in *Pinus ponderosa* under -8 Bar and in *Pinus elderica* under -12 Bar water stress levels.

Semerci et al. [5] studied the influence of water stress on various *Pinus nigra* origins. They created -2, -4, and -6 Bar water stress levels. They analyzed the seeds' germination percentages based on the origins. They concluded that germination percentage varied between the origins, and increasing water stress led to a significant reduction in germination percentages. The study included seeds originating in Ballıköy in Tavsanlı. The germination percentage of the seeds in the control group was 98%. This rate reduced to 76% under -2 Bar, 52% under -4 Bar, and 16% under -6 Bar water stress levels. However, the seeds originating in Göksun B. Çamurlu had a germination percentage of 62% in the control group, while this rate was 58% for Andırın Akifiye-origin seeds. The germination was at a rate of 1% under -2 Bar water stress level for these seeds, while no germination took place under -6 Bar water stress level.

The same study also reported that proportional germination percentages largely vary between the origins. For instance, proportional germination percentage in seeds originating in Tavsanlı Ballıköy was 77% under -2 Bar, 53% under -4 Bar, and 17% under -6 Bar water stress levels. Similarly, proportional germination percentage in seeds originating in Mengen Daren was 75% under -2 Bar, 63% under -4 Bar, and 16% under -6 Bar water stress levels. However, the same study revealed that the proportional germination percentage of seeds originating in Göksun B. Çamurlu and Andırın Akifiye was only 2% under -2 Bar water stress level. Topacoglu et al. [91] reported that *Pinus nigra* originating in Ankara Uluhan yielded 95.08% cumulative germination percentage under -8.0 Bar water stress level, while the seed originating in Isparta Tota yielded a cumulative germination percentage of 85.41%.

Buyurukçu [27] compared the tolerance of Anatolian pinus nigra (*Pinus nigra* Arnold. subsp. *pallasiana* (Lamb.) Holmboe) clones from clonal seed garden in Hanönü Günlüburun against the drought. In order to compare the clones against drought during the phase of germination, the seeds were subjected to -2, -4, -6, and -8 Bar water stress levels using PEG 6000 solution.

All in all, it was seen at the end of ANOVA and Duncan's test results that clones had different responses of tolerance against water stress. Germination percentage, which was 48% on average in the control group, reduced to 16% under -2 Bar, 15% under -4 Bar, 2% under -6 Bar, and 0.4% under -8 Bar. However, what is remarkable in this study is the great differences between clones. For instance, the 17th clone had a germination percentage of 29% in the control group, while this rate reduced to 2% under -2 Bar water stress level. The seed did not germinate under the higher water stress levels. Similarly, the 9th clone had a germination percentage of 78% which reduced to 0% under -2 Bar water stress level. However, the 14th clone had a germination percentage of 49%, which reduced to 8% under -6 Bar water stress conditions and 5% under -8 water stress conditions. Taking into account the fact that these clones belong to the same stand of origin and are grown under the same conditions, it is possible to say that genetic structure may be prominent in determining the tolerance towards water stress.

5. Recommendations

The plants used for parks and recreational areas consist of a wide variety. Previous studies report that water stress works in which PEG solution is used yield successful results. It is possible to have an idea about the water stress tolerance of alternative species to be used in an area where landscape work will be practiced through a 1-month preliminary work performed in this area. Thus, it will be possible to use species which are tolerant to water stress. However, if researchers conduct such studies and provide the practitioners with ready-to-use information, this will bring a great ease for them. Therefore, such studies should be varied and increased in number.

Due to global warming, drought and scarcity of water make their devastating effects felt in all aspects of our lives including green areas. Therefore, the water should be used reasonably, which makes drought-tolerant plants and their practices crucial. The plants to be used in landscape areas are required to be drought-tolerant. It sometimes functions as a crucial criterion determining plant choice in some areas. In order to use water reasonably, the species and origins which are tolerant to drought should be determined. Using the seeds and saplings of such species and origins in landscape arrangements is of great importance. In this sense, species should be subjected to drought stress under equal conditions and compared. Accordingly, selecting the species which are most tolerant to drought for landscape practices is important.

The fact that species have rather variable tolerance against water stress has been revealed in previous studies. Even the regions and territories where water scarcity has not influenced the growth of tree species much yet may experience problems due to drought stress in near future. Hence, it is even important to compare the drought stress tolerances of populations in the same climatic regions. Therefore, the situation of local origins is not guarantees, which makes the identification of relatively tolerant species important as part of decision-making on future forestation strategies.

Therefore, it is very important to perform experiments on the origins of the species that have been proved to be drought-tolerant in order to determine their most drought-tolerant origins through studies focusing on extreme fields and to produce seeds from them. Hence, these seeds can be used for landscape practices, which may yield great benefits in future. The use of such species, particularly in areas where maintenance will be performed after mining activities, areas prone to erosion, areas involving forestation on side slopes and traffic islands, urban forests in arid areas, and so on, will both reduce watering, maintenance, fertilization, etc. costs and extend the lifespan of saplings.

Author details

Nurcan Yigit[1], Hakan Sevik[2], Mehmet Cetin[3*] and Nur Kaya[4]

*Address all correspondence to: mehmet.cetin@temple.edu

1 Faculty of Forestry, Department of Forestry, Kastamonu University, Kastamonu, Turkey

2 Faculty of Engineering and Architecture, Department of Environmental Engineering, Kastamonu University, Kastamonu, Turkey

3 Faculty of Engineering and Architecture, Department of Landscape Architecture, Kastamonu University, Kastamonu, Turkey

4 Institute of Science, Kastamonu University, Kuzeykent, Kastamonu, Turkey

References

[1] Korkmaz K: The effect of global warming and agricultural practices. Alatarım, 2007; 6 (2): 43–49. (in Turkish).

[2] Kaya N: The effect of drought stress on the germination percentages in some tree species, Kastamonu University. Institute of Science and Technology. Department of Forestry Engineering, M.Sc., 2014: 64 pp. Kastamonu (in Turkish).

[3] Şevik H, Karakaş H, Şenöz E: Evaluation of air quality in terms of the amount of carbon dioxide in Black Sea Region. International Journal of Engineering Sciences & Research, 2013; 2 (2): 173–177.

[4] Özden Ş, Tetik Ç, Yavaş ÖM, İlgen HG, Çiftçi A: Europe needs to be done to work on climate change adaptation and disaster reduction of damage due to climate change in Turkey. 5th World Water Forum regional preparatory process DSi domestic regional water meetings snow hydrology conference proceedings, Erzurum, 2008; pp. 95–103. (in Turkish).

[5] Semerci H, Öztürk H, Semerci A, İzbırak A, Ekmekçi Y: Larch in different reclamation zones sampled Anatolia (*Pinus nigra* Arnold. ssp. nigra var. caramanica (loudon)

Rehder) to determine the origin of frost and drought tolerance. Ankara: Agenda offset. Ministry of Environment and Forestry Publication No: 340 Forest Trees and Seeds Research Institute Technical Bulletin, 21. ISBN 978-605-393-017-4. 2008. (in Turkish).

[6] Kulaç Ş: Drought stress exposed Scots pine (*Pinus sylvestris L.*) seedlings in some morphological and physiological Investigation of biochemical changes. Ph.D. Thesis, Karadeniz Technical University, Institute of Science and Technology. 2010. Trabzon. (in Turkish).

[7] Yaldız G, Sekeroğlu N: The importance of medicinal and aromatic plants in the global climate change. Journal of Turkish Scientific Assembly, 2013; 6 (1): 85–88.

[8] Demir I, Mavi K: Effect of salt and osmotic stresses on the germination of pepper seeds of different maturation stages, Brazilian Archives of Biology and Technology, 2008; 51 (5): 897–902.

[9] Ertop G: Global Warning and Xeriscape. M.Sc. Thesis, Ankara University, Institute of Science, Ankara, 2009; 144p. (In English with abstract)

[10] Pratap V, Sharma YK: Impact of osmotic stress on seed germination and seedling growth in black germ (*Phaseolus mungo*). Journal Environ Biology, 2010: 31 (5): 721–726.

[11] Karaca E, Kusvuran A: The evaluation of water-wise landscape of certain plants which were used on Cankiri landscape arrangements. Turkish Journal of Scientific Reviews, 2012; 5 (2): 19–24, (In English with abstract)

[12] Yilmaz H, Yilmaz H: The examining of usage areas of naturally growing woody plants in highway slopes: Erzurum-Uzundere case, Süleyman Demirel University. Journal of Forestry Faculty, 2009; A (1); 101–111.

[13] Raziuddin Swati ZA, Bakht J, Farhatullah Ullah N, Shafi M, Akmal M, Hassan G: In situ assessment of morpho-physiological response of wheat (*Triticum aestivum* L.) genotypes to drought. Pakistan Journal of Botany, 2010; 42 (5): 3183–3195.

[14] Almas DE, Bagherikia S, Mashaki KM: Effects of salt and water stresses on Germination and seedling growth of *Artemisia vulgaries* L. International Journal of Agriculture and Crop Sciences, 2013; 6 (11): 762–765.

[15] Hamayun M, Khan SA, Shinwari ZK, Khan A, Ahmad N, Lee IJ: Effect of polyethylene glycol induced drought stress on physio-hormonal attributes of soybean. Pakistan Journal of Botany, 2010; 42 (2): 977–986.

[16] Ialelou FS, Shafagh-Kolvanagh J, Fateh M: Effect of salinity on germination indexes of medicinal plant naked pumpkin (*Cucurbita pepo*). International Journal of Agriculture and Crop Sciences, 2013; 5 (13): 1424–1426.

[17] Muhammed Z, Hussain F: Efect of NaCl salinity on the germination and seedling growth of some medicinal plants. Pakistan Botany Journal, 2010; 42 (2): 889–897.

[18] Shitole SM, Dhumal KN: Effect of Water stressby polyethylene glycol 6000 and sodium chloride on seed germination and seedling growth of *Cassia angustofolia*. International Journal of Pharma Sciences and Research 2012; 3 (2): 528–53.

[19] Sayar R, Bchini H, Mosbahi M, Khemira H: Response of durum wheat (*Triticum durum* desf.) growth to salt and drought stresses. Czech Journal of Genetics Plant Breeding, 2010; 46 (2): 54–63.

[20] Ahmad S, Ahmad R, Ashraf MY, Ashraf M, Waraich, EA: Sunflower (Helianthus annuus l.) response to drought stress at germination and seedling growth stages. Pakistan Journal of Botany, 2009; 41 (2): 647–654.

[21] Mohammadizad HA, Khazaei I, Ghafari M, Sinehsar MFF, Barzegar R: Effect of salt and drought stresses on seed germination and early seedling growth of nepeta persica. International Journal of Farming and Allied Sciences, 2013; 2 (21): 895–899.

[22] Bahrami H, Razmjoo J, Ostadi JA: Effect of drought stress on germination and seedling growth of sesame cultivars (*Sesamum indicum* L.). International Journal of Agricultural Sciences, 2012; 2: 423–428.

[23] Mujtaba SM, Ali M, Ashraf MY, Khanzada B, Farhan SM, Shirazi MU, Khan MA, Shereen A, Mumtaz S: Physiological responses of wheat (*Triticum aestivum* L.) Genotypes under water stress conditions at seedling stage. Pakistan Journal of Botany, 2007; 39 (7): 2575–2579.

[24] Michel BE, Kaufmann MR: The osmotic potential of polyethylene glycol 6000. Plant Physiology, 1973; 51 (5): 914–916.

[25] Meneses CHSG, Bruno RLA, Fernandes PD, Pereira WE, Lima LHGM, Lima MMA, Vidal MS: Germination of cotton cultivar seeds under water stress induced by polyethyleneglycol-6000. Sci. Agric. (Piracicaba, Braz.). Crop Science, 2011; 68 (2): 131–138.

[26] Özdamar K: Data analysis using SPSS statistical package programs with MINITAP. Eskisehir: Fourth Edition, Carroll Bookstore. 1999.

[27] Buyurukçu S: Hanönü-Günlübur the Anatolian black pine (*Pinus nigra* Arnold ssp. *Pallasiana Lamb.* Holmboe) Water Stress Effects on Seed Garden in Terms of clonal variation. Master Thesis, Kastamonu University, Institute of Science and Technology. Kastamonu, 2011. (in Turkish).

[28] Khera N, Shing RP: Germination of some multipurpose tree species in five provenances in response to variation in light temperature, substrate and water stress. Tropical Ecology, 2005; 46 (2): 203–217.

[29] Öztürk S, Bozdoğan E: Determination of the perceived quality of urban life in New and traditional housing textures. Fresenius Environmental Bulletin, 2014; 23 (10): 2415–2421.

[30] Papinchak H, Holcomb EJ, Orendovici BT, Decoteau, DR: Effectiveness of houseplants in reducing the indoor air pollutant ozone. HortTechnology, 2009; 19 (2): 286–290.

[31] Tani A, Hewitt CN: Uptake of aldehydes and ketones at typical indoor concentrations by houseplants. Environmental Science and Technology, 2009; 43 (21): 8338–8343.

[32] Cetin M: Chapter 55: Using Recycling Materials for Sustainable Landscape Planning. Environment and Ecology at the Beginning of 21st Century. Ed.: Prof. Dr. Recep Efe, Prof. Dr. Carmen Bizzarri, Prof. Dr. İsa Cürebal, Prof. Dr. Gulnara N. Nyusupova, ST. Kliment Ohridski University PRESS, Sofia; 2015. pp. 783–788, ISBN: 978-954-07-3999-1.

[33] Öztürk S, Bozdoğan E: The contribution of urban road trees on improving the air quality in an urban area. Fresenius Environmental Bulletin, 2015; 24 (5a): 1822–1829.

[34] Sevik H, Cetin M, Belkayali N: Effects of forests on amounts of CO_2: case study of Kastamonu and Ilgaz Mountain National Parks. Polish Journal of Environmental Studies, 2015; 24 (1): 253–256.

[35] Yiğit N, Öztürk A, Sevik H: Ecological impact of urban forests (Example of Kastamonu urban forest). International Journal of Engineering Sciences & Research Technology, 2014; 3 (12): 558–562.

[36] Sevik H, Belkayalı N, Aktar G: Change of chlorophyll amount in some landscape plants. Journal of Biotechnological Sciences, 2014; 2 (1): 10–16.

[37] Belkayali N. Kesimoglu MD: The stakeholders' point of view about the impact of recreational and tourism activities on natural protected area: a case study from Kure Mountains National Park, Turkey, Biotechnology & Biotechnological Equipment, 2015; 29 (6): 1092–1103.

[38] Cetin M: Evaluation of the sustainable tourism potential of a protected area for landscape planning: a case study of the ancient city of Pompeipolis in Kastamonu. International Journal of Sustainable Development & World Ecology, 2015; 22 (6): 490–495. doi:10.1080/13504509.2015.1081651

[39] Cetin M: Determining the bioclimatic comfort in Kastamonu City. Environmental Monitoring and Assessment, 2015; 187 (10): 640. doi:10.1007/s10661-015-4861-3

[40] Cetin M: Using GIS analysis to assess urban green space in terms of accessibility: case study in Kutahya. International Journal of Sustainable Development & World Ecology, 2015; 22 (5): 420–424. doi:10.1080/13504509.2015.1061066

[41] Turna I, Guney D: Altitudinal variation of some morphological characters of Scots pine (*Pinus sylvestris* L.) in Turkey. African Journal of Biotechnology, 2009; 8 (2): 202–208.

[42] Tekçe N, Belkayali N, Oğuz D, Bastemur CT: A survey on recreational use of domestic water supply reservoirs: a case study from Kurtboğazi–Ankara, Turkey. African Journal of Agricultural Research, 2010; 5 (14): 1897–1907

[43] Talay İ, Kaya F, Belkayalı N: Socio-economic structure effects on the recreational trends and demands: a case study Bartin city. Journal of Geographical Sciences, 2010; 8 (2): 147–156. (in Turkish).

[44] Djukanovic R, Wargocki P, Fanger P O: Cost–benefit analysis of improved air quality in an office building. Proceedings: Indoor Air, 2002; pp. 808–813, Monterey, California, USA

[45] Lohr VI, Pearson-Mims CH, Goodwin GK: Interior plants may improve worker productivity and reduce stress in a windowless environment. Journal of Environmental Horticultural, 1996; 14: 97–100.

[46] Chang C, Chen P: Human response to window views and indoor plants in the workplace. HortScience, 2005; 40: 1354–1359.

[47] Cetin M, Sevik H: Measuring the impact of select plants on indoor CO2 concentrations, Polish Journal of Environmental Studies, 2016. 25(3), 1–7, http://www.pjoes.com/doi/61744.pdf

[48] Geven F, Bingöl Ü, Güney K: The Revision of Polygonaceae Family at Herbarium ANK and to Prepare of the Database. Kastamonu University, Journal of Forestry Faculty, 2008; 8 (1): 67–85.

[49] Cetin M: A change in the amount of CO2 at the center of examination halls. Studies on Etno–Medicine, 2016, 10(2), 146–155

[50] Guney K, Cetin M, Sevik H, Guney KB: Influence of germination percentage and morphological properties of some hormones practice on Lilium martagon L. seeds. Oxidation Communication, 2016, 39(1–I), 466–474

[51] Topaçoğlu O, Bozkus HF, GÜNEY K: Silvicultural characteristics of some stand structures in the subalpine and high montane altitudinal zones on the north aspect of Ilgaz Mountain. Kastamonu University, Journal of Forestry Faculty, 2008; 8 (1): 1–13.

[52] Šálek L, Güney K: New host plant for the species *Agapanthia lateralis* Gangl.(Coleoptera; Cerambycidae). Entomological News, 2014; 124 (1): 29–32.

[53] Ozcan GE, Eroglu M, Akinci HA: Use of pheromone-baited traps for monitoring Ips sexdentatus (Boerner)(Coleoptera: Curculionidae) in oriental spruce stands. African Journal of Biotechnology, 2013; 10 (72): 16351–16360.

[54] Guloglu Y: Legal Aspects of Urban Forests in Turkey, Session G1: Governance, 16th European forum on urban Forestry, 7-11 May 2013, Milano, Italy.

[55] Oğuz D, Diriöz ED, Belkayalı N: Tourists' perception of landscape design: The case of resorts in the Belek specially protected area. African Journal of Agricultural Research, 2010; 5 (10): 1028–1035.

[56] Güloğlu Y, Belkayalı N, Sevik H: Legal Analysis of Plant Species Protection in Turkey, Proceedings of Kyrgyz First International Congress of Biology, Biskek/Kırgızistan, September 2012; pp. 88, 24–27

[57] Belkayali N, Sevik H, Guloglu Y: What affects perceptions of local residents toward protected areas? A case study from Kure Mountains National Park, Turkey. International Journal of Sustainable Development & World Ecology, 2016; 23 (2): 194–202. doi: 10.1080/13504509.2015.1098743

[58] Aricak B, Kucuk O, Enez K: Determining a fire potential map based on stand age, stand closure and tree species, using satellite imagery (Kastamonu central forest directorate sample). Croatian Journal of Forest Engineering, 2014; 35 (1): 101–108

[59] Ozcan GE, Cicek O, Enez K, Yildiz M: A new approach to determine the capture conditions of bark beetles in pheromone-baited traps. Biotechnology & Biotechnological Equipment, 2014; 28 (6): 1057–1064.

[60] Gumus S, Aricak B, Enez K, Acar HH: Analysis of tree damage caused by rockfall at forest road construction works. Croatian Journal of Forest Engineering, 2009; 30 (2): 151–158.

[61] Sevik H, Guney K: Effects of some hormone applications on morphological features of *Melissa officinalis* L. root cuttings. Soil–Water Journal, 2013; 2 (2): 1647–1652

[62] Sevik H, Guney K: Effects of IAA, IBA, NAA, and GA3 on rooting and morphological features of *Melissa officinalis* L. stem cuttings. The Scientific World Journal, 2013; 2013:5. Article ID 909507.

[63] Aricak B: Using remote sensing data to predict road fill areas and areas affected by fill erosion with planned forest road construction: a case study in Kastamonu Regional Forest Directorate (Turkey). Environmental Monitoring and Assessment, 2015; 187 (7): 1–10.

[64] Güney D, Atar F, Atar E, Turna I, Kulaç S: The effect of pre-treatments and seed collection time on the germination characteristics of common hornbeam (*Carpinus betulus*) seeds in the Eastern Black Sea Region, Turkey. Seed Science and Technology, 2015; 43 (1): 1–9.

[65] Cetin M: Chapter 27: Landscape Engineering, Protecting Soil, and Runoff Storm Water. Advances in Landscape Architecture, InTech, Ed: Murat Özyavuz, 2013, pp. 697–723. doi:10.5772/55812

[66] Bingöl Ü, Geven F, Güney K: Heavy metal (Pb and Ni) accumulation in the branch and bark tissues of street tree *Sophora japonica* L. Orman Fakültesi Dergisi, 2008; 8 (1): 93–96.

[67] Kesik HI, Korkut S, Hiziroglu S, Sevik H: An evaluation of properties of four heat treated wood species. Industrial Crops and Products, 2014; 60: 60–65.

[68] Yiğit N, Öztürk A, Sevik H: Ecological impact of urban forests (example of Kastamonu urban forest). International Journal of Engineering Sciences & Research Technology, 2014; 3 (12): 558–562

[69] Öztürk S, Bilgili B: The use of Natural plant species of Kastamonu University campus in Landscape. Kastamonu University Journal of Forestry Faculty, 2015; 15 (1): 175–182.

[70] Sevik H, Topacoglu O: Variation and inheritance pattern in Cone and seed characteristics of Scots pine (*Pinus sylvestris* L.) for evaluation of genetic diversity. Journal of Environmental Biology, 2015; 36 (5): 1125–1130

[71] Turna I, Güney D: Altitudinal variation of some morphological characters of Scots pine (*Pinus sylvestris* L.) in Turkey. African Journal of Biotechnology, 2009; 8 (2): 202–208.

[72] Guney D, Yahyaoglu Z, Turna I, Mueller-Starck G: Genetic variation in *Pinus brutia* in Turkey. Fresenius Environmental Bulletin, 2014; 23 (5): 1249–1254.

[73] Akinci H, Ozcan G, Eroglu M: Impacts of site effects on losses of oriental spruce during Dendroct onus micans (Kug.) outbreaks in Turkey. African Journal of Biotechnology, 2009; 8(16), 3934–3939

[74] Canbulat, S: Checklist of Turkish Raphidioptera on the basis of distribution pattern and biogeographical analysis. Turkish Journal of Zoology, 2015; 39 (2): 225–234

[75] Kaya LG, Cetin M, Doygun H: A holistic approach in analyzing the landscape potential: Porsuk Dam Lake and its environs, Turkey. Fresenius Environmental Bulletin, 2009; 18 (8): 1525–1533.

[76] Canbulat S: Redescription of Osmylus multiguttatus McLachlan, 1870 (Neuroptera: Osmylidae) with distributional remarks, ZOOTAXA, 2013; 3741 (3): 385–390.

[77] Canbulat S: Two new records of the Coniopteryx (Neuroptera: Coniopterygidae) species for the Turkish fauna. Turkish Journal of Zoology, 2013; 37 (2): 242–245

[78] Cetin M: Consideration of permeable pavement in landscape architecture. Journal of Environmental Protection and Ecology, 2015; 16 (1): 385–392.

[79] Cetin M, Sevik H: Evaluating the recreation potential of Ilgaz Mountain National Park in Turkey. Environmental Monitoring and Assessment, 2016; 188 (1): 52. doi:10.1007/s10661-015-5064-7

[80] Guloglu Y: The establisment of forest ownership and the legal regulations on the forests until the Tanzimat (reform) period in the Ottoman State. Kastamonu University, Journal of Forestry Faculty, 2015; 10 (2): 180–194.

[81] Guloglu Y: Evaluation of forest land allocation for tourism with regard to agents of administrative act. Kastamonu University, Journl of Economics and Administrative Sciences, 2015; 10: 45–55.

[82] Ozturk S, Enez K: Determination of the perceptions of secondary education students towards environment and nature. Journal of Environmental Protection and Ecology, 2015; 16 (2): 723–732.

[83] Özden S, Tetik Ç, Yavas, ÖM, İlgen HG, Çiftçi A: Europe needs to be done to work on climate change adaptation and disaster reduction of damage due to climate change in Turkey. 5th World Water Forum regional preparatory process DSi domestic regional water meetings proceedings of snow hydrology conference, Erzurum, 2008; pp. 95–103.

[84] Sevik H, Cetin M: Effects of water stress on seed germination for select landscape plants. Polish Journal of Environment Studies, 2015; 24 (2): 689–693.

[85] Falusi M, Calamassi R, Tocci A: Sensitivity of seed germination and seedling root growth to moisture stress in four provenances of *Pinus halepensis* Mill. Silvea Genetica, 1993; 32 (1–2): 4–9.

[86] Tilki F, Dirik H: Seed germination of three provenances of *Pinus brutia* (Ten.) as influenced by stratification, temperature and water stress. Journal of Environmental Biology, 2007; 28 (1): 133–139.

[87] Boydak M, Dirik H, Tilki F, Çalıkoğlu M: Effects of water stress on germination in six provenances of *Pinus brutia* seeds from different bioclimatic zones in Turkey. Turkish Journal Agriculture and Forestry, 2003; 27: 91–97.

[88] Ahmadloo F, Tabari M, Behtari B: Soil and erosion. International Journal of Forest, 2011; 1 (1): 11–17.

[89] Kaufmann M, Eckard A: Water potential and temperature effects on germination of engelmann spruce and lodgepole pine seeds. Forest Science, 1977; 23 (1), 27–33.

[90] Djavanshir K, Reid CPP: Effect of moisture stress on germination and radicle development of *Pinus eldarica* Medw. and *Pinus ponderosa* Laws. Canadian Journal of Forest Research, 1975; 5 (1): 80–83.

[91] Topacoglu O, Sevik H, Akkuzu E: Effects of water stress on germination of *Pinus nigra* Arnold. Seeds. Pakistan Journal of Botany, 2016; 48(2), 447–453

Chapter 4

Ameliorating Drought-Induced Stress in Turfgrass through Genetic Manipulation

Mostafa K. Sarmast

Additional information is available at the end of the chapter

http://dx.doi.org/10.5772/64201

Abstract

To delineate the major processes associated with short water scarcity in four tall fescue species, we examined their enzymatic and nonenzymatic antioxidant activity and *FaSGR* expression levels. Moreover, we examined the possibility of *Agrobacterium*-mediated transformation of *Arabidopsis P5CS1* gene in tall fescue. According to the results, proline has been introduced as an important compatible osmolyte, so as to protect enzymes and cellular structures under water scarcity. In addition to that, superoxide dismutase (SOD) along with proline can be used as a core physiological indicator for the assessment of adaptability to environmental conditions. Results indicated that most of the superoxide that was produced as a result of drought stress was converted to H_2O_2 by SOD and subsequently detoxified by ascorbate peroxidase (APX) into H_2O. Notably, the *FaSGR* transcript increased drastically over the course of the drought stress in Pixie and Mini-mustang, in contrast to jaguar and h–d, supporting the notion of Stay GReen (*SGR*)-mediated chlorophyll degradation in the less drought-tolerant cultivars. Different modulations of ROSs quenching system in tall fescue genotypes suggest that even one stress signal causes different signaling responses in different cultivars. The heterologous transformation of *P5CS1* in *Festuca arundinacea* background, confirmed by PCR and transient *GUS* assay, most probably can improve tall fescue tolerance to drought stress.

Keywords: drought stress, enzymes, proline, *SGR*, tall fescue

1. Introduction

1.1. Drought stress

A world in which average temperatures reach 4°C above preindustrial levels would likely see unprecedented heat waves, severe drought, major flooding, and up to 1 m of sea-level rise (http://

INTECH
open science | open minds

© 2016 The Author(s). Licensee InTech. This chapter is distributed under the terms of the Creative Commons Attribution License (http://creativecommons.org/licenses/by/3.0), which permits unrestricted use, distribution, and reproduction in any medium, provided the original work is properly cited.

www.worldbank.org/). The aforementioned change in environment clearly is because of the careless human activities. In nature, plants as sessile organisms must respond to different stimuli throughout their life cycle; otherwise, fluctuations of these abiotic stresses outside their normal ranges can cause serious consequences on plant growth and reproduction. Many physiological and biochemical processes will be changed in plants under stress by affecting RNA stability, protein, and ion transport [1].

Stress is a specific environmental condition herein plant growth and development disrupted and its full genetic potential defeated [1]. Up to 70% of food crops yield is negatively influenced by biotic stresses, such as drought, cold, high salinity, and heat, which obviously threaten food security worldwide. Many organic compounds such as proline, glycine betaine, and polyamines, a variety of sugars (mainly fructose and sucrose), sugar alcohols, complex sugars (such as trehalose and fructans), and organic acids (oxalate, malate) can be aggregated inside the cell, while plants have faced with salt, drought, and cold to prevent cellular dehydration and protect cellular proteins. These osmoprotectants presumably accumulate several folds more than unstressed condition, without disturbing the intracellular biochemistry. Drought stress is by far the most important environmental factor contributing to yield losses in crops [2]. Plants experienced drought stress either when the water supply to roots becomes difficult or when the transpiration rate becomes very high. These two conditions often coincide under arid and semiarid climates. Imbalance between light capture and its utilization caused by drought stress inhibits photosynthetic activity [3]. The changes in the photochemistry of chloroplasts due to fluctuation of electrons in the leaves of drought-stressed plants result in the dissipation of excess light energy in the PSII core and antenna, thus generating reactive oxygen species (ROS), which are potentially dangerous under drought stress conditions [4]. Due to multigenic and quantitative properties of aforementioned stresses, advanced plant breeding purposes have been challengeable. High throughput sequencing and functional genomics tools provide new window to underlie signal perception and transduction of the molecular regulatory networks. The mechanism by which plants perceive and transmit the signals down to active adaptive response inside the cell is of great importance to mimic this strategy for manipulating organelle to convey tolerance properties under stress conditions. To do this, merging physiological, biochemical, and gene regulatory network knowledge is inevitable [5]. The objectives of the present work were as follows:

1. Elucidating cellular enzymatic and nonenzymatic antioxidative defense strategies in the drought stress response in leaves of four genotypically distinct *F. arundinacea* Schreb. genotypes during midsummer

2. Studying the regulatory role of *NYE*/Stay GReen (*SGR*) transcript during drought stress

3. Overexpressing of the *P5CS1* gene in tall fescue background aimed at improving drought tolerance of this species

1.2. Turfgrasses importance

The *Poaceae* family comprise over 9000 monocot species in which environmental grasses distribution is pretty diverse. Perennial grass has a specific role in providing food for livestock,

http://dx.doi.org/10.5772/64201

bioenergy, fiber products, soil and water improvement and conservation, habitats for wildlife population, and recreation and beautification [6]. Over 50 grass species are cultivated as turfgrass, which have a major role in our daily life by providing a ground cover in parks, home lawns, sports fields, and golf courses [7]. Turfgrasses are also used along roadsides, edges of waterways, or in preserved land, where they may act as barriers to reduce pollution, water runoff, and wind erosion of the underlying soil [7]. The turfgrass industry potentially is a multibillion-dollar-a-year business. Turfgrass provide attractive vegetative ground cover, which is consistently mowed.

Turf species are divided into two groups based on origin and geographical distribution: cool-season grasses and warm-season grasses. Cool-season turf species are typically adapted to the cool–humid and cool–arid zones, and warm-season turf species are best adapted to the warm–arid and warm–humid regions. Most of the cool-season grasses are propagated by seeds, while a majority of the warm-season grasses are vegetatively propagated. Maintaining the quality of turf under adverse environmental conditions is of prime importance and actually a big challenge in turf growth and production. This challenge could be addressed through molecular breeding to create stress-tolerant germplasm. Traditional breeding techniques resulted in the production of many genotypes with an improved stress tolerance; however, traditional breeding progress is limited mostly due to the lack of superior stress-tolerant grass germplasm and because of poor understanding of the different aspects of physiological and molecular mechanisms for stress tolerance in perennial grass species. Finding and overexpressing of some drought-related candidate genes, either transcription factors, key enzymes, or channel proteins, recognized by RNA seq and other high-throughput techniques resulted in improving plant adaptation to drought stress [8]. The molecular mechanism and genetic information underlying drought stress have not yet been well addressed in turfgrass species.

1.3. Origin, taxonomy, and cytological features of *Festuca arundinacea* Schreb.

An increasing interest in tall fescue (*Festuca arundinacea* Schreb.) in western Europe and elsewhere is mainly because of its better drought resistance and yield potential in contrast to that of perennial ryegrass (*Lolium perenne* L.) [9]. Tall fescue is native to Europe; it also grows on the Baltic coasts throughout the Caucasus, in western Siberia, and extends into China [10].

The genus *Festuca* encompasses over 80 species, including two agriculturally important forage crops *Festuca arundinacea*, which is one of the most widely used cool-season species, and *F. pratensis* Huds [11]. Tall fescue is a wind-pollinated species with a high degree of self-incompatibility. Its genome size is approximately 6×10^3 Mbp (mega base pair) and contains three genomes (P, G1, and G2). The P (2*x*) genome is from the diploid species, meadow fescue (*F. pratensis*), while the G1 and G2 (4*x*) genomes are from the tetraploid species, *F. arundinacea* var. "glaucescens" [12, 13]. In comparison to other cool-season perennial grasses, tall fescue exhibits a high degree of stability when confronting drought stress. In contrast to having low rate of water use, tall fescue's deep and expansive root system allows this species to avoid drought conditions by continually having access to water [14]; even under continuous dry periods, the cells remain turgid. Transpiration is another strategy for tall fescue in order to cool itself via vast leaves [15]. This relative drought tolerance makes it an ideal option for cultivation in urban landscapes throughout transitional climates. Tall fescue is a coarse-textured vigorous perennial bunchgrass. This species reproduces through tillering and seed. Its desirable agronomic characteristics encompass high yields of herbage, excellent persistence, adaptation to a wide range of soil conditions, compatibility with various management practices, long grazing season, and low incidence of pest problems [16, 17]. Given the continuous drought stress in many areas of the world, the necessity of highly tolerant turf species for landscaping, xeriscaping, and other usage is inevitable. Selection and/or production of high-tolerant transgenic turfgrass would be one of the key ways that leads to ameliorate their drought tolerance threshold in the realm of landscaping.

2. Drought stress and experimental design

Two replicate experiments with potted *F. arundinacea* were conducted. The treatment for each genotype had four replicates (four pots). The control plants were maintained in an optimal soil water condition (FC) during the whole experimental period. The experiment was carried out in loamy soil that was collected at a height of 20 cm from the top of the department's research field in plastic pots (45 cm in height). Potted plants grown under greenhouse conditions were subjected to drought by withholding irrigation for 8 days, whereupon leaves withered and started to discolor in most genotypes. Mini-mustang, Pixie, Jaguar, and h–d were selected in this experiment to capture the range of genetic and phenotypic (aesthetic value) diversity observed in the 11 cultivars. Fully expanded youngest mature leaves of the aforementioned genotypes were collected at days 2, 4, 6, and 8 after the start of the drought treatment for biochemical experiments. Well watered pots (irrigated daily) served as a control for collecting leaf tissue. Leaf samples were collected at midday and were immediately frozen in liquid nitrogen and stored at −80°C until analysis.

2.1. Biochemical experiment

2.1.1. Enzyme assay

The methods used to determine the activities of catalases (CAT) and ascorbate peroxidase (APX) were those reported by Ref. [18]. Superoxide dismutase (SOD) activity was estimated by measuring the decrease in absorbance of superoxide nitro blue tetrazolium complex by the enzyme [19]. The activity of peroxidase (POX) was determined using the method of Ref. [20], with minor changes. Proline content was quantified using the protocol described by Ref. [21].

2.2. qRT-PCR analysis

Total Ribonucleic acid (RNA) extraction, cDNA synthesis, and qRT-PCR analysis were performed as described previously [22].

2.3. Tall fescue tissue culture and callus regeneration

Tall fescue (*F. arundinacea*) seeds were submerged into 25% sulfuric acid for 30 min and then prewashed in tap water overnight. These seeds were moved into the airflow cabinet hood; then they were soaked into 25–50% Clorox (containing 5.25% sodium hypochlorite) solution containing 0.02% household detergent for 30 min for surface sterilization and then rinsed six times with sterilized distilled water. Finally, the dehusked seeds were cut longitudinally and placed onto MS [23] basal medium supplemented with 0–16 mg/l 2,4-Dichlorophenoxyacetic acid (2,4-D), in which explants gained the ability of callus induction at dark in 2 weeks. These calli were subcultured consequently for more than a year, every 4 weeks, on MS media containing 5–8 mg/l 2,4-D. We tried to keep the embryogenic calli in subcultures which were identifiable easily by the yellowish color and firm texture compared to the nonembryogenic calli. In all these experiments, the calli were kept at dark. BAP (0.1, 0.5 mg/l) and kinetin (0.1 and 0.2 mg/l), W/O 2,4-D, have been considered for plant regeneration of tall fescue calli. The pH of all media was adjusted to 5.8 by 0.1 N HCl before autoclaving for 15 min at 121°C and 1.5 kg/cm^2 pressure. Cultures were kept at 25 ± 2°C temperature under cool white fluorescent light (30 μmol/m^2/s), with 16/8 h day/night photoperiods.

2.4. *Agrobacterium*-mediated transformation

pGV3101 strain of *Agrobacterium tumefaciens* harboring pBI121 plasmid as a binary vector contains the *gus* gene under the control of the cauliflower mosaic virus' (CaMV) 35S promoter and the selectable marker neomycin phosphotransferase II (nptII) under the control of the CaMV promoter. Aforesaid bacteria were grown and selected in rotator (200 rpm) in LB liquid media (tryptone 10 g/l, NaCl 5 g/l, yeast extract 5 g/l) supplemented with 50 μg/ml kanamycin for 24 h at 28°C. The cells were harvested by centrifugation (8000 rpm for 10 min at 4°C) and further resuspended in 10–15 ml of MS medium. Acetosyringone (AS) was added to the medium up to 100 μM. For transformation, 4 × 4 mm of calli grown at dark were considered for inoculation with *A. tumefaciens* (OD_{600} = 0.5–1) under 400 mg Hg pressure for 10–15 min; then, the callus pieces and *Agrobacterium* were incubated together for 20 min with gentle shaking. Excess bacteria were removed after the incubation. The infected callus pieces were transferred onto filter papers for a few minutes and then placed onto cocultivation MS medium supplemented with 100 μM acetosyringone (AS) and 5 mg/l 2,4-D for almost 3 days (in dark) at 25°C. Sometimes, we used to use an empty petri dish containing a Whatman filter paper, moisturized with sterile water, for cocultivation. After cocultivation, the explants were washed with sterile distilled water once; for the second time, the explants were incubated with either 500 mg/l cefotaxime or 400 mg/l Timentin for about 30 min, along with gentle shaking to prevent *Agrobacterium* overgrowth later on (in regeneration MS media); then they were blotted onto sterilized filter papers and placed on MS medium supplemented with 5–8 mg/l 2,4-D, 150–200 mg/l G418 for the selection of transformed explants and 200 mg/l Timentin to prevent *Agrobacterium* overgrowth for 2 weeks. For plant regeneration, explants were transferred to regeneration media supplemented either with BA (6-Benzylaminopurine) or kinetin in light. Cultures were kept at 25 ± 3 °C under cool white fluorescent light (30 μm/m^2/s), for 16 h each day.

2.4.1. *PCR confirmation and gus histochemical staining*

The presence of *uidA* gene in embryogenic calli was confirmed by PCR amplification of tall fescue-calli genomic DNA with gene-specific primers: *gus*-F (GCTGTGCCAGGCAGTTT-TAAC) and *gus*-R (ATATCGTCCACCCAGGTGTTC). The predicted size of the amplified DNA fragments of *uidA* was 425 bp. DNA amplifications were performed in a total volume of 20 μl containing 1 μl of 10 μM forward pimer, 1 μl of 10 μM reverse primer, 2 μl of 10× Ex Taq Buffer, 0.5 μl of dNTP mixture (2.5 mM each), 0.1 μl of TakaRa Ex Taq enzyme (5 unit/μl) (Takara, Shuzo, Kyoto, Japan), by Thermocycler (Bio-Rad, United States). PCR was carried out for screening of regenerated transformed plantlets with an initial denaturation at 94°C for 5 min, followed by 30 cycles of 94°C, 30 min; 58°C, 30 min; 72°C, 30 s; and a final extension, 75°C for 10 min. The PCR products were separated in 1% agarose gel containing 0.5 μg/ml ethidium bromide. The size of the amplification products was estimated using a 100 bp DNA ladder (GeneRuler DNA Ladder Mix # SM 0331, Fermentas). GUS histochemical staining was performed as described previously [24].

3. Tall fescue biochemical response to short drought stress

In response to water stress, plants exert adaptive modifications in their morphological, physiological, and biochemical properties. Photosystem II (PSII) is more vulnerable than PSI during drought stress in which inhibition of CO_2 assimilation, coupled with the changes in photosystem activities and photosynthetic electron transport capacity, results in accelerated production of reactive oxygen species (ROS). The damaged targets by ROS attack are recovered by repair or by replacement via de novo biosynthesis. However, under severe drought stress, cell death would be inevitable. The fate of stressed cells is determined by the duration of stress as well as the protective capacity of the plant. Reactive oxygen species play a crucial role in causing cellular damage under drought stress [25]. Reactive oxygen species not only play a signaling role in coordinating nuclear gene expression in order to protect cells during biotic and abiotic stress responses, but can also cause lipid peroxidation and consequently membrane injury, protein degradation, and enzyme inactivation [26, 27]. All plants have ROS detoxification mechanisms, which are enzymatic, with superoxide dismutase (SOD), catalase (CAT), ascorbate peroxidase (APX), peroxidase (POD), glutathione reductase (GR), and monodehydro ascorbate reductase (MDAR), and nonenzymatic detoxification mechanisms through flavanones, anthocyanins, carotenoids, and ascorbic acid (AA). The activity of antioxidant enzymes under drought stress will be enormously flexible among several plant species and even between two cultivars of one species. It has been demonstrated that enzymatic breakdown of ROS is one of the major processes the plant uses to scavenge this signaling molecule [28]. Superoxide dismutase, the first enzyme in the detoxifying process, converts superoxide anion radicals ($O^{2?}$) to hydrogen peroxide (H_2O_2), and APX reduces H_2O_2 to water using ascorbic acid as a specific electron donor [29–31].

It has been stated that ROS not only cause damage to membrane, but also disturb the correct functions of DNA and proteins. Enzyme-mediated disruption of ROS effectively quenches ROS in plant cells [28].

To underlie the key processes concomitant with drought tolerance in tall fescue, four genotypes (Jaguar, h–d, Pixie, and Mini-mustang) were assessed for enzymatic and nonenzymatic antioxidant activity as well as *SGR* expression during water shortage. These cultivars were selected from a larger set of 11 tall fescue cultivars based on the genetic relationships delineated via intersimple sequence repeat (ISSR) analysis [22]. These cultivars were thus selected for preliminary experiments in which drought stress was induced by ceasing irrigation in plastic pots for up to 8 days. Two antioxidant enzymes' activity (markers of drought stress) increased in all four cultivars due to drought stress—APX and SOD. Increased SOD activity of Mini-mustang, Pixie, jaguar, and h–d (6 days only) is in agreement with [32], who observed similar patterns when heat stress was applied on two annual and two perennial cool-season turfgrasses. A strong correlation between SOD activity and drought-induced oxidative stress tolerance has been well established [33]. At day 8, SOD activity slightly declined in h–d. Such decline in SOD activity in severe drought stress has been previously reported [34]. Chlorophyll stability was associated with leaf APX activity in the Jaguar and h–d leaves. APX is known as a chlorophyll protector in leaves [35, 36]. The connection between SOD and APX activity in most genotypes suggests that most of the superoxide that was produced as a result of drought stress was converted to H_2O_2 and subsequently detoxified by APX into H_2O [37]. In transgenic tall fescue, high levels of APX and SOD were considered as defense mechanisms during stress conditions [38]. Regardless of Jaguar, all cultivars showed a nonsignificant decline in CAT activity at 8 days. Such decline in CAT activity has been reported by several authors in different stress conditions [33, 34, 39, 40]. Pixie and h–d genotypes experienced a nonsignificant decline in catalase activity, which does not seem to be in accordance with chlorophyll changes which is in line with the reports of [41, 42]. Identification of some chloroplast proteins that potentially can be involved in the regulation of drought stress in *F. arundinacea* has been reported by [43] during short water scarcity in tall fescue. In another report, the higher photosynthetic capacity of an intergeneric hybrid of *L. multiflorum*/*F. arundinacea* was likely due to higher efficiency of Calvin cycle during the drought stress [44]. The kinetic variation observed among the antioxidant activity most likely depends on species and cultivars, time and type of stress, tissue or organ types, as well as severity of stress [28, 45].

4. Osmoregulatory effects of proline

Proline is a very specific amino acid essential for primary metabolism, which can accumulate in eubacteria, protozoa, marine invertebrates, and plants under stress condition. First proline accumulation report has been addressed in welting perennial ryegrass [46]. Soon after that, numerous reports have shown proline accumulation under drought [47], high salinity [48], high light and ultraviolet (UV) irradiation [49], heavy metal [50], oxidative stress [32], and biotic stress [51], in which proline metabolism by far has mainly been studied in response to osmotic stress [52]. Proline is synthesized in cytosol mainly from glutamate, which can convert to proline in two steps. First, glutamate is reduced to glutamate-semialdehyde (GSA) and then spontaneously converted to pyrroline-5-carboxylate (P_5C) by the pyrroline-5-carboxylate synthetase (P_5CS) and P_5C reductase (P_5CR), respectively [53, 54]. Proline is eventually catabolized to P_5C in mitochondria by proline dehydrogenase (PDH) or proline oxidase, and then P_5C dehydrogenase (P_5CDH) converts P_5C to glutamate [55]. Kumar et al. [56] evaluated T_1 transgenic plants of *indica* rice overexpressing a P_5CS gene. Their research indicated a better growth performance, biomass production, higher proline accumulation, and lower rate of lipid peroxidation in transgenic plants compared to nontransgenic plants under 150 mM NaCl stress. Furthermore, the ameliorating effects of proline on heavy metal stress have been reported on *Chlamydomonas reinhardtii*, which was able to express P_5CS gene 80% higher than wild type [57].

Proline can be encoded by P_5CS_1 and P_5CS_2 in cytosol. While P_5CS_1 translational fusion of GFP normally localizes in cytosol, but right after osmotic stress, the P_5CS_1 signal was detectable in chloroplast but not P_5CS_2. Székely et al. [58] and Strizhov et al. [59] reported the augmentation of proline under osmotic stress in chloroplast, which was induced by P_5CS. This report sheds light on different subcellular compartments of proline biosynthesis, which is pretty different based on the environmental conditions. A proline uniporter, which facilitates proline transport into the mitochondrial matrix, and a proline/glutamate antiporter, which appears to have an important role in the Pro/Glu shuttle between the mitochondrial matrix and the cytosol, have been identified in *Triticum durum* Desf. mitochondria [60]. In *Arabidopsis*, P_5CS_2 is a house-keeping gene, whereas *Arabidopsis* P_5CS_1 is induced by osmotic and salt stresses [48]. Proline has been shown to function as a molecular chaperone able to protect protein integrity and enhance the activities of different enzymes. Proline acts as a singlet oxygen quencher and has a ROS scavenging activity. Hien et al. [61] reported that proline accumulation in *Oryza sativa* was related to induction of proline biosynthesis by osmotic stress and not proteolysis. In *Opuntia* spp. exposed to heat, salinity, and water-deficit stress, an increase in the proline content was reported, but this accumulation was not necessarily correlated with P5CS enzyme activity, whereas the transcript level of *P5CS* was correlated with proline accumulation in the report of Silva-Ortega et al. [62]. We evaluated the proline content of the tall fescue leaf during the course of drought stress, because predominately proline accumulation is a physiological plant response to biotic and abiotic stress. The research results indicated that proline content in all tested genotypes increased significantly, especially at the highest deficit irrigation treatment. For example, Pixie showed over 50-fold increase in proline content than control plants, while this was about twofold increase in h–d genotype. We concluded that the substantial increase in proline content in Pixie and a lesser extent in Mini-mustang likely are because of their dependency on proline, so as to maintain cell homeostasis during water stress. However, other drought-tolerance mechanisms may be responsible for slight changes of proline in h–d and Jaguar. Our research results collectively suggest that proline is an important compatible osmolyte which serves as a protectant for enzymes and cellular structures of tall fescue under severe drought stress.

5. Overexpression of *Arabidopsis P5CS* in *F. arundinacea* calli

Particle bombardment–genetic transformation of *F. rubra* L. was one of the preliminary reports by Ref. [63]. However, 12 years later, the first successful *F. arundinacea Agrobacterium*-mediated transformation was published [64]. Since the success of *Agrobacterium*-mediated transformation of rice in the early 1990s, transformations of some other monocotyledonous species occurred routinely. *Agrobacterium*-mediated transformation has several advantages over the biolistic method, not the least of which is the stable integration and expression of the target gene in offspring, mostly due to lower copy number, and fewer rearrangements in genome [47, 65]. Bettany et al. [64] were able to regenerate only one transgenic *F. arundinacea* expressing *uidA* gene after establishment in soil, even though they failed to produce seeds. The *Agrobacterium* LBA4404 was the strain that had been used in their research. They reported a huge variability in their transgenic events, which in most cases did not follow Mendelian segregation [64].

Soon after the last aforementioned report, Wang and Gee [66] developed a much more efficient *Agrobacterium*-mediated transformation procedure aimed at regenerating transgenic tall fescue fertile plants. They reported that the number of hygromycin-resistant calli obtained per dish was in the range of 2.0–5.8; the number of transgenic plants recovered per dish was in the range of 0.4–1.7, and a 1:1 segregation ratio of the transgenes was found in the progenies. Gao et al. [67] gained 10.5% average transformation efficiency across the four callus lines of *F. arundinacea*. In their report, *Agrobacterium*-mediated transformation appears to be the preferred method for producing transgenic tall fescue plants. Zhao et al. [68] examined whether salt tolerance can be improved stably by overexpressing vacuolar Na^{+}/H^{+} (*AtNHX1*) antiporters in tall fescue. Their research leads to identifying a single-copy inheritance of $AtNHX_1$—performing better in the presence of 200 mM NaCl than control plants—in most of the T1 and T2 lines of tall fescue after *Agrobacterium*-mediated transformation, with a near 1:1 segregation ratio which subsequently has been approved for release by the Chinese Department of Agriculture [68]. Dong and Qu [69] reported that *Agrobacterium*-mediated transformation of tall fescue yields 34% *hyg* B-resistant calli and had 8% overall transformation efficiency. Hu et al. [53] had improved the cold resistance of tall fescue through *Agrobacterium* transformation of *ipt* gene. Transgenic plants have had higher chlorophyll content and stayed greener, besides having a higher tillering ability as in contrast to that of control plant, which is of great economic importance to improve the tolerance of this plant to environmental stress. To the best of our knowledge, the *Agrobacterium*-mediated transformation of *F. arundinacea* by *P5CS* gene has not yet been reported.

The finding that abiotic stress is involved in chloroplast–stroma protein degradation under drought stress in numerous plant species, together with the demonstration of enhanced osmotic stress tolerance in plants by proline overproduction, spurred us on applying P_5CS gene encoding proline for improving drought stress in *F. arundinacea*. The preliminary work on callus production resulted in a high amount of callus during a year and so, with different types of calli. Soft, yellowish calli were chosen as a primary sample for agrotransformation. Whereas the callus regeneration rate was astonishing, plant regeneration was far behind callus regeneration. An increase in regeneration efficiency was observed in media supplemented with 8 mg/l 2,4-D, which was almost six times greater than 2 mg/l 2,4-D treatment by dehusked, longitudinally sliced seeds. Apart from 2,4-D concentration, dehusked *Festuca* mature seeds showed a further callus induction efficiently as opposed to intact seeds. The callus regeneration frequency from small *Festuca* callus explants was also enhanced in dark as compared with light situation. Acetosyringone in cocultivation media greatly improved *Agrobacterium* growth, while acetosyringone-free MS media delayed *Agrobacterium* growth and subsequently, most probably, could decrease *Agrobacterium*-mediated transformation efficiencies. The heterologous transformation of *P5CS* in *F. arundinacea* background was confirmed by PCR and transient *GUS* assay, which most probably can improve tall fescue tolerance more to drought stress.

6. NYE/SGR protein role during drought-mediated chlorophyll degradation

During senescence, plant-recycled valuable nutrient components from the leaves and leaf chlorophyll (Chl) are usually converted to a colorless product through six known Chl catabolism enzymes (CCEs) and a metal-chelating substance (MCS) [70]. First, Chl *b* reductase (CBR) reduces 7-formyl group of Chl *b* to a hydroxymethyl group. There are two CBR isoforms which are encoded by *NON-YELLOW COLORING 1* (*NYC1*) and *NYC-LIKE* (*NOL*) genes. Another key regulator of Chl degradation is hydroxymethyl Chl reductase (HCAR), which reduces Chl *b* to Chl *a* [71, 72]. HCAR has recently been identified in *Arabidopsis*. Then, the Mg^{+} atom of Chl *a* is removed by MCS, which is then called pheophytin *a* (Phein *a*). Pheophytinase (PPH) produces Pheophorbide *a* (Pheide *a*) by catalyzing Phein *a*. Subsequently, the chlorin macrocycle of Pheide *a* is oxygenolytically opened by Pheide *a* oxygenase (PAO) [73], and red Chl catabolite (RCC) which is the product of this reaction is reduced to a nonphototoxic primary fluorescent Chl catabolite (*p*FCC) by RCC reductase (RCCR). Besides CCEs and MCS, STAY-GREEN1 (SGR1) or nonyellowing mutant can cause a stay-green phenotype in many plant species, such as *Arabidopsis* and rice [74]. The *NYE1* (nonyellowing) gene in *Arabidopsis* has been identified by positional cloning. NYE1 is now widely referred as SGR (Stay GReen) [75]. The *nye1-1* mutant could retain 50% chlorophyll at the end of 6-day dark incubation, whereas the wild type (*Columbia-0*) has resulted in the degradation of chlorophyll to less than 10% [75]. In addition to that, qPCR result has outlined *AtNYE1* as an extremely induced gene by senescence signals [75].

In *Festuca arundinacea*, this gene is called NONYELLOWING PROTEIN1 [nonyellowing gene (NYE)], which has 278 amino acids [76]. *FaNYE1* or *SGR1* has a high similarity to *Arabidopsis NYE1*, either by sequences or by function. *FaNYE* has been identified recently by RACE-PCR [76]. Overexpression of *AtNYE1* results in pale-yellow leaves to even albino seedlings [74]. Degreening phenotype in tall fescue occur during severe stress conditions and harsh seasons. BLAST analysis revealed that *FaNYE* has 89% sequence similarity to *Triticum urartu* Tumanian ex Gandilyan and 83% sequence similarity to *Hordeum vulgare* (NCBI). The negative correlation of *FaNYE* transcript with chlorophyll content has been previously addressed in *F. arundinacea* affected by dark treatment and natural senescence, which in 9 days' dark incubation augmented *FaNYE* transcript level just about 52-fold. Furthermore, overexpressing *FaNYE* ORF in *Col-0* background leads to accelerated senescence [76]. Leaf chlorophyll concentration of the mutant diminished very slowly in rice *sgr* mutant, but steeply decreased in wild type during the grain filling. However, no difference in photosynthetic activity was observed between the stay green mutant and the yellowing wild-type leaves [77].

We hypothesized that the SGR-mediated chlorophyll degradation in fescue is genetically dependent on *FaSGR*. To confirm this, we evaluated the *FaSGR* transcripts in our selected genotypes during drought stress by real-time PCR. The substantial increase in *FaSGR* transcript in Pixie and Mini-mustang cultivars suggests their SGR-mediated chlorophyll degradation. The *FaNYE1/SGR* role in chlorophyll degradation during dark treatment and natural senescence has been published [76]. During stress conditions, chloroplast metabolism modulates leaf senescence. Hence, prevention of chlorophyll degradation can be an interesting subject in order to generate new varieties of plants with higher performance. In a preliminary transient experiment on tall fescue leaves, we targeted tall fescue *SGR* via CRISPR-Cas9 technology under heat and salinity stress, and then realized that the leaves that were exposed to CRISPR-Cas9 construct vividly maintained more chlorophyll than control lines.

7. Conclusion

The kinetic results of this research noticeably suggest that proline, SOD, and APX probably can be key components to protect tall fescue cells from severe drought stress. Furthermore, the aforementioned components can be key targets for genetic manipulation of tall fescue, so as to maintain its cell homeostasis against drought stress. In addition, proline and SOD can be considered as the main physiological indicators for the assessment of adaptability to environment conditions. Also, targeting SGR in leaves of tall fescue can open a new window for future research on leaf senescence.

Author details

Mostafa K. Sarmast

Address all correspondence to: mkhsarmast@gau.ac.ir

Department of Horticultural Science, Faculty of Plant Production, Gorgan University of Agricultural Sciences and Natural Resources, Gorgan, Golestan, Iran

References

[1] Taiz L, Zeiger E, Møller IM, Murphy A. Plant Physiology and Development. 6th ed. Sinauer Associates, Inc., Canada. 2015; p. 761.

[2] Weber RLM, Wiebke-Strohm B, Bredemeier C, Margis-Pinheiro M, de Brito GG, Rechenmacher C, Bertagnolli PF, de Sá MEL, Campos MA, de Amorim RMS, Beneventi MA, Margis R, Grossi-de-Sa MF, Bodanese-Zanettini MH. Expression of an osmotin-like protein from *Solanum nigrum* confers drought tolerance in transgenic soybean. BMC Plant Biol. 2014; 14: 343.

[3] Foyer CH, Noctor G. Oxygen processing in photosynthesis: Regulation and signaling. New Phytol. 2000; 146: 359–388.

[4] Peltzer D, Dreyer E, Polle A. Temperature dependencies of antioxidative enzymes in two contrasting species. Plant Physiol. Biochem. 2002; 40: 141–150.

[5] Thakur P, Kumar S, Malik JA, Berger JD, Nayyar H. Cold stress effects on reproductive development in grain crops: An overview. Environ. Exp. Bot. 2010; 67: 429–443.

[6] Lewandowski I, Scurlockb JMO, Lindvallc E, Christou M. The development and current status of perennial rhizomatous grasses as energy crops in the US and Europe. Biomass Bioenerg. 2003; 25: 335–361.

[7] Turgeon AL. Turfgrass Management. 8th ed. Prentice-Hall, Englewood. NJ. 2008.

[8] Bhatnagar-Mathur P, Vadez V, Sharma KK. Transgenic approaches for abiotic stress tolerance in plants: Retrospect and prospects. Plant Cell Rep. 2008; 27: 411–424.

[9] Cougnon M, Baert J, Van Waes C, Reheul D. Performance and quality of tall fescue (*Festuca arundinacea* Schreb.) and perennial ryegrass (*Lolium perenne* L.) and mixtures of both species grown with or without white clover (*Trifolium repens* L.) under cutting management. Grass Forage Sci. 2013; 69: 666–677.

[10] Jauhar PP. Cytogenetics of the *Festuca-Loliu*m Complex: Relevance to Breeding. Springer, Berlin. 1993.

[11] Yamada T. Festuca. In: Kole C (ed.). Wild Crop Relatives: Genomic and Breeding Resources. Springer, New York. 2011; p. 318.

[12] Seal AG. DNA variation in Festuca. Heredity 1983; 50: 225–236.

[13] Sleper D.A. Breeding tall fescue. Plant Breed. Rev. 1985; 3: 313–342.

[14] Carrow RN. Drought avoidance characteristics of diverse tall fescue cultivars. Crop Sci. 1996; 36: 371–377.

[15] Cross JW, Bonos SA. Huang B, Meyer WA. Evaluation of heat and drought as components of summer stress on tall fescue genotypes. HortScience 2013; 48: 1562–1567.

[16] Hanson AA. The future of tall fescue. In: Buckner RC, Bush LP (eds.). Tall Fescue. American Society of Agronomy, Crop Science Society of America, Soil Science Society of America, Madison. 1979; pp. 341–344.

[17] Sleper DA, West CP. Tall fescue. In: Moser LE, Buxton DR, Casler MD (eds.). Cool-season Forage Grasses. American Society of Agronomy; Crop Science Society of America; Soil Science Society of America, Madison. 1996; pp. 471–502.

[18] Jiang M, Zhang J. Involvement of plasma-membrane NADPH oxidase in abscisic acid and water stress-induced antioxidant defense in leaves of maize seedlings. Planta 2002; 215:1022–1030.

[19] Dhindsa RS, Plumb-Dhindsa P, Thorpe TA. Leaf senescence: Correlated with increased levels of membrane permeability and lipid peroxidation, and decreased levels of superoxide dismutase and catalase. J. Exp. Bot. 1981; 32: 93–101.

[20] Chance B, Maehly A. Assay of catalase and peroxidase. In: Colowick SP, Kaplan NO (eds.). Methods in enzymology. Academic Press, New York, 1995; pp. 764–775.

[21] Bates LS, Waldern RP, Teare ID. Rapid determination of free proline for water stress study. Plant Soil. 1973; 39: 205–207.

[22] Sarmast MK, Salehi H, Niazi A. Biochemical differences underlie varying drought tolerance in four *Festuca arundinacea* Schreb. genotypes subjected to short water scarcity. Acta Physiol. Plant. 2015; 37: 192.

[23] Murashige T, Skoog F. A revised medium for rapid growth and bioassays with tobacco tissue culture. Physiol. Plant. 1962; 15: 473–497.

[24] Jefferson RA. Assaying chimeric genes in plants: The GUS gene fusion system. Plant Mol. Biol. Rep. 1987; 5: 387–405.

[25] Reddy AR, Chaitanyaa KV, Vivekanandan M. Drought-induced responses of photosynthesis and antioxidant metabolism in higher plants. J. Plant Physiol. 2004; 161: 1189–1202.

[26] Sairam RK, Srivastava GC, Saxena DC. Increased antioxidant activity under elevated temperatures, a mechanism of heat stress tolerance in wheat genotypes. Biol. Plant. 2000; 43: 245–251.

[27] Maruta T, Noshi M, Tanouchi A, Tamoi M, Yabuta Y, Yoshimura K, Ishikawa T, Shigeoka S. H2O2-triggered retrograde signaling from chloroplasts to nucleus plays specific role in response to stress. J. Biol. Chem. 2012; 287: 11717–11729.

[28] Huang B, DaCosta M, Jiang Y. Research advances in mechanisms of turfgrass tolerance to abiotic stresses: From physiology to molecular biology. Crit. Rev. Plant Sci. 2014, 33: 141–189.

[29] Asada K. Ascorbate peroxidase–a hydrogen peroxide scavenging enzyme in plants. Physiol. Plant. 1992; 85: 235–241.

[30] Asada K. The water-water cycle in chloroplasts: Scavenging of active oxygens and dissipation of excess photons. Ann. Rev. Plant Physiol. 1999; 50: 601–639.

[31] Foyer CH, Noctor G. Redox signaling in plants. Antioxid. Redox Signal. 2013;18: 2087–2090.

[32] Yang Z, Miao Y, Yu J, Liu J, Huang B. Differential growth and physiological responses to heat stress between two annual and two perennial cool-season turfgrasses. Sci. Hort. 2014; 170: 75–81.

[33] Pan Y, Wu LJ, Yu ZL. Effect of salt and drought stress on antioxidant enzymes activities and SOD isoenzymes of liquorice (*Glycyrrhiza uralensis* Fisch). Plant Growth Regul. 2006; 49: 157–165.

[34] Liu J, Xie X, Du J, Sun J, Bai J. Effects of simultaneous drought and heat stress on Kentucky bluegrass. Sci. Hort. 2008; 115: 190–195.

[35] Sofo A, Dichio B, Xiloyannis C, Masia A. Antioxidant defences in olive trees during drought stress: Changes in activity of some antioxidant enzymes. Func. Plant Biol. 2005a; 32: 45–53.

[36] Sofo A, Tuzio AC, Dichio B, Xiloyannis C. Influence of water deficit and rewatering on the components of the ascorbate-glutathione cycle in four interspecific *Pronus* hybrids. Plant Sci. 2005b; 162: 403–412.

[37] Mhamdi A, Queval G, Chaouch S, Vanderauwera S, Van Breusegem F, Noctor G. Catalase function in plants: A focus on Arabidopsis mutants as stress-mimic models. J. Exp. Bot. 2010; 61: 4197–4220.

[38] Lee SH, Ahsan N, Lee KW, Kim DH, Lee DG, Kwak SS, Kwon SY, Kim TH, Lee BH. Simultaneous overexpression of both CuZn superoxide dismutase and ascorbate peroxidase in transgenic tall fescue plants confers increased tolerance to a wide range of abiotic stresses J. Plant Physiol. 2007; 164: 1626–1638.

[39] Feierabend J, Engel S. Photoinactivation of catalase in vitro and in leaves. Arch. Biochem. Biophys. 1986; 251: 567–576.

[40] Polle A. Defense against Photooxidative Damage in Plants [A]. Oxidative Stress and the Molecular Biology of Antioxidant Defense [C] Harbor. Cold Spring Harbor Laboratory Press, Cold Spring, NY. 1997; pp. 785–813.

[41] Sharma P, Dubey RS. Modulation of nitrate reductase activity in rice seedlings under aluminium toxicity and water stress: Role of osmolytes as enzyme protectant. J. Plant Physiol. 2005; 162: 854–864.

[42] Simova-Stoilova L, Vaseva I, Grigorova B, Demirevska K, Feller U. Proteolytic activity and cysteine protease expression in wheat leaves under severe soil drought and recovery. Plant Physiol. Biochem. 2010; 48: 200–206.

[43] Kosmala A, Perlikowski D, Pawlowicz Rapacz M. Changes in the chloroplast proteome following water deficit and subsequent watering in a high- and a low-drought-tolerant genotype of Festuca arundinacea. J Exp Bot 2012; 63: 6161–6172.

[44] Perlikowski D, Kosmala A, Rapacz M, Koscielniak J, Pawlowicz Zwierzykowski Z. Influence of short-term drought conditions and subsequent re-watering on the physiology and proteome of Lolium multiflorum/Festuca arundinacea introgression forms, with contrasting levels of tolerance to long-term drought. Plant Biol. 2013; 16: 385–394.

[45] Fu J, Huang B. Involvement of antioxidants and lipid peroxidation in the adaptation of two cool-season grasses to localized drought stress. Environ. Exp. Bot. 2001; 45: 105–114.

[46] Kemble AR, MacPherson HT. Liberation of amino acids in perennial ray grass during wilting. Biochem. J. 1954; 58: 46–59.

[47] Choudhary NL, Sairam RK, Tyagi A. Expression of delta1-pyrroline-5-carboxylate synthetase gene during drought in rice (*Oryza sativa* L.). Ind. J. Biochem. Biophys. 2005; 42: 366–370.

[48] Yoshiba Y, Kiyosue T, Katagiri T, Ueda H, Mizoguchi T, Yamaguchi-Shinozaki K, Wada K, Harada Y, Shinozaki K. Correlation between the induction of a gene for Δ^1-pyrroline-5-carboxylate synthetase and the accumulation of proline in *Arabidopsis thaliana* under osmotic stress. Plant J. 1995; 7: 751–760.

[49] Saradhi P, Alia P, Arora S, Prasad KV. Proline accumulates in plants exposed to UV radiation and protects them against UV induced peroxidation. Biochem. Biophys. Res. Commun. 1995; 209: 1–5.

[50] Schat H. Heavy metal-induced accumulation of free proline in a metal-tolerant and a nontolerant ecotype of *Silene vulgaris*. Physiol. Plant. 1997; 101: 477–482.

[51] Haudecoeur E, Planamente S, Cirou A, Tannières M, Shelp BJ, Moréra S, Faure D. Proline antagonizes GABA-induced quenching of quorum-sensing in *Agrobacterium tumefaciens*. PNAS. 2009; 106: 14587–14592.

[52] Verbruggen N, Hermans C. Proline accumulation in plants: A review Amino Acids 2008; 35: 753–759.

[53] Hu CAA, Delauney AJ, Verma DPS. A bifunctional enzyme (Δ1-pyrroline-5-carboxylate synthetase) catalyzes the first two steps in proline biosynthesis in plants. PNAS 1992; 89: 9354–9358.

[54] Szoke A, Miao GH, Hong Z, Verma DP. Subcellular location of delta-pyrroline-5-carboxylate reductase in root/nodule and leaf of soybean. Plant Physiol. 1992; 99: 1642–1649.

[55] Szabados L, Savoure A. Proline: A multifunctional amino acid. Cell 2009; 15: 89–97.

[56] Kumar V, Shriram V, Kavi Kishor PB, Jawali M, Shitole MG. Enhanced proline accumulation and salt stress tolerance of transgenic indica rice by over-expressing *P5CSF129A* gene. Plant Biotechnol. Rep. 2010; 4: 37–48.

[57] Siripornadulsi S, Traina S, Verma DPS, Sayre RTS. Molecular mechanisms of proline-mediated tolerance to toxic heavy metals in transgenic microalgae. Plant Cell 2002; 14: 2837–2847.

[58] Székely G, Abrahám E, Cséplo A, Rigó G, Zsigmond L, Csiszár J, Ayaydin F, Strizhov N, Jásik J, Schmelzer E, Koncz C, Szabados L. Duplicated *P5CS* genes of Arabidopsis play distinct roles in stress regulation and developmental control of proline biosynthesis. Plant J. 2008; 53: 11–28.

[59] Strizhov N, Abrahám E, Okrész L, Blickling S, Zilberstein A, Schell J, Koncz C, Szabados L. Differential expression of two *P5CS* genes controlling proline accumulation during salt-stress requires ABA and is regulated by *ABA1, ABI1* and *AXR2* in Arabidopsis. Plant J. 1997; 12: 557–569.

[60] Di Martino C, Pizzuto R, Pallotta ML, De Santis A, Passarella S. Mitochondrial transport in proline catabolism in plants: The existence of two separate translocators in mitochondria isolated from durum wheat seedlings. Planta 2006; 223: 1123–1133.

[61] Hien DT, Jacobs M, Angenon G, Hermans C, Thu TT, Son LV, Roosens NHR. Proline accumulation and D1-pyrroline-5-carboxylate synthetase gene properties in three rice cultivars differing in salinity and drought tolerance. Plant Sci. 2003; 165: 1059–1068.

[62] Silva-Ortega CO, Ochoa-Alfaro AE, Reyes-Agüero JA, Aguado-Santacruz GA, Jiménez-Bremont JF. Salt stress increases the expression of *P5CS* gene and induces proline accumulation in cactus pear. Plant Physiol. Biochem. 2007; 46: 82–89.

[63] Ha SB, Wu FS, Thorne TK. Transgenic turf-type tall fescue (*Festuca arundinacea* Schreb.) plants regenerated from protoplasts. Plant Cell Rep. 1992; 11: 601–604.

[64] Bettany AJE, Dalton SJ, Timms E, Manderyck B, Dhanoa MS, Morris P. *Agrobacterium tumefaciens*-mediated transformation of *Festuca arundinacea* (Schreb.) and *Lolium multiflorum* (Lam.). Plant Cell Rep. 2003; 21: 437–444.

[65] Dai S, Zheng P, Marmey P, Zhang S, Tian W, Chen S, Beachy RN, Fauquet C. Comparative analysis of transgenic rice plants obtained by *Agrobacterium* mediated transformation and particle bombardment. Mol. Breed. 2001; 7: 25–33.

[66] Wang ZY, Gee Y. Agrobacterium-mediated high efficiency transformation of tall fescue (*Festuca arundinacea*). J. Plant Physiol. 2004; 162: 103–113.

[67] Gao C, Long D, Lenk I, Nielsen KK. Comparative analysis of transgenic tall fescue (*Festuca arundinacea* Schreb.) plants obtained by *Agrobacterium*-mediated transformation and particle bombardment Plant Cell Rep. 2008; 27: 1601–1609.

[68] Zhao J, Zhi D, Xue Z, Liu H, Xia G. Enhanced salt tolerance of transgenic progeny of tall fescue (*Festuca arundinacea*) expressing a vacuolar Na+/H+ antiporter gene from *Arabidopsis*. J. Plant Physiol. 2007; 164: 1377–1383.

[69] Dong S, Qu R. High efficiency transformation of tall fescue with *Agrobacterium tumefaciens*. Plant Sci. 2005; 168: 1453–1458.

[70] Hortensteiner S, Krutler B. Chlorophyll breakdown in higher plants. Biochim. Biophys. Acta 2011; 1807: 977–988.

[71] Horie Y, Ito H, Kusaba M, Tanaka R, Tanaka A. Participation of chlorophyll b reductase in the initial step of the degradation of light-harvesting chlorophyll *a*/*b*-protein complexes in Arabidopsis. J. Biol. Chem. 2009; 284: 17449–17456.

[72] Sakuraba Y, Kim YS, Yoo SC, Hortensteiner S, Paek, NC. 7-Hydroxymethyl chlorophyll a reductase functions in metabolic channeling of chlorophyll breakdown intermediates during leaf senescence. Biochem. Biophys. Res. Commun. 2013; 430: 32–37.

[73] Pružinsk A, Tanner G, Anders I, Roca, Hortensteiner S. Chlorophyll breakdown: Pheophorbide an oxygenase is a Rieske-type iron-sulfur protein, encoded by the accelerated cell death 1 gene. PNAS 2003; 100: 15259–15264.

[74] Park SY, Yu JW, Park JS, Li J, Yoo SC, Lee NY, Lee SK, Jeong SW, Seo HS, Koh HJ, Jeon JS, Park YI, Paek NC. The senescence induced stay green protein regulates chlorophyll degradation. Plant Cell 2007; 19: 1649–1664.

[75] Ren G, An K, Liao Y, Zhou X, Cao Y, Zhao H, Ge X, Kuai B. Identification of a novel chloroplast protein AtNYE1 regulating chlorophyll degradation during leaf senescence in arabidopsis. Plant Physiol. 2007; 144: 1429–1441.

[76] Wei Q, Guo Y, Kuai B. Isolation and characterization of a chlorophyll degradation regulatory gene from tall fescue. Plant Cell Rep. 2011; 30: 1201–1207.

[77] Cha KW, Lee YJ, Koh HJ, Lee BM, Nam YW, Paek NC. Isolation, characterization, and mapping of the stay green mutant in rice. Theor. Appl. Genet. 2002; 104: 526–532.

Chapter 5

Water Use Strategy of Four Desert Shrubs in Gonghe Basin, Qinghai-Tibetan Plateau

Yajuan Zhu

Additional information is available at the end of the chapter

http://dx.doi.org/10.5772/63195

Abstract

Gonghe basin is located in the ecotone from the semi-arid to arid zone on the northeastern Qinghai-Tibetan Plateau. *Caragana intermedia* and *Caragana korshinskii* are dominant on sand dunes. *Salix psammophila* and *Salix cheilophila* are mainly distributed on interdune. Water source of four desert shrubs was examined by δD and $\delta^{18}O$, and their long-term water use efficiency (WUE) was compared by leaf $\delta^{13}C$. Four desert shrubs used different depths of soil water depending on their availability in different seasons, including shallow soil water recharged by rain in spring or summer and deep soil water recharged by ground water. The reliability on ground water of two *Salix* shrubs on interdune was more significant than two *Caragana* shrubs on sand dunes. Moreover, the WUE of four shrubs decreased in drought spring. Two shrubs in *Caragana* had similar WUE in the growing season. However, the WUE of *S. psammophila* was always higher than that of *S. cheilophila*, which might have more adaptability in Gonghe Basin.

Keywords: water source, water use efficiency (WUE), stable isotope, soil water, ground water

1. Introduction

In desert ecosystems, water is a restrictive factor for plant survival and growth because of low and unpredictable precipitation and high evaporation [1, 2]. The ability to use rainwater in

INTECH open science | open minds

© 2016 The Author(s). Licensee InTech. This chapter is distributed under the terms of the Creative Commons Attribution License (http://creativecommons.org/licenses/by/3.0), which permits unrestricted use, distribution, and reproduction in any medium, provided the original work is properly cited.

spring and summer is important for plant phenology and growth [3]. Additionally, the sustainable water source is necessary for plant growth especially in the drought period, such as deep soil water or ground water [4]. A stable isotope technology is often used to study the water use strategy of desert plants. Generally, there is no stable isotope fractionation during water uptake by root system or water transportation in the xylem of most plant species. Thus, the main water source can be distinguished by comparing the δD or $\delta^{18}O$ value of xylem water with that of the potential water source, for example, rain, snow, river, lake, soil water or ground water [1]. Further, leaf $\delta^{13}C$ value of C_3 plants is positively related to their long-term water-use efficiency (WUE). The $\delta^{13}C$ value decreased from spring to autumn in the growing season and increased in the drought period [4].

Previous studies with the stable isotope of hydrogen or oxygen indicated that woody plant species used different water sources in desert or other arid and semi-arid ecosystems. Firstly, many trees and shrubs mainly used shallow soil water recharged by rain, for example, *Artemisia ordosica* in Mu Us Sandy Land [5], *Pinus edulis* and *Juniperus osteosperma* in Utah [6], *Larix sibirica* and *Potentilla fruticosa* in the mountain forest along the Kherlen River of Mongolia [7], *Senecio filaginoides* and *Mulinum spinosum* in Patagonian steppe [8], *Nitraria tangutorum* and *Artemisia arenaria* in Badain Jaran Desert [9], *Sarcobatus vermiculatus* in San Luis Valley of Colorado [10] and *Haloxylon persicum* in Gurbantunggut Desert [11]. Secondly, some deep-rooted shrubs and trees mainly used ground water, for example, *Sabina vulgaris* and *Salix matsudana* in Mu Us Sandy Land [12], *Populus fremontii* in the non-saline habitat of Colorado River Basin [13], *Ericameria nauseosa* in San Luis Valley [10], *Ulmus pumila* in Hunshandake Sandy Land [14], *Pinus sylvestris* var. *mongolica* in Horqin Sandy Land [15], *Populus euphratica* along Heihe River [16] and *Haloxylon ammodendron* in Gurbantunggut Desert [11]. Thirdly, few shrubs or trees with dimorphic root systems used both soil water recharged by rain and deep soil water recharged by ground water and even ground water. *S. vermiculatus* used ground water during dry periods but used deep soil water large amounts of rainfall in San Luis Valley [10].

Gonghe Basin is located in the northeastern Qinghai-Tibetan Plateau, which is the ecotone from semi-arid to arid region, with the altitude varying from 2600 to 3400 m. It contains part of Gonghe County, Guinan County and Xinghai County of Qinghai Province. The dominant vegetation is steppe and desert steppe. The basin is as long as 210 km in East and West, and as wide as 60 km in South and North, with a total area of 13,800 km^2. It is one of the most vulnerable land desertification area in Qinghai Province. The area of desertified land is 3530 km^2, which accounts for 25.58% of the total area. The ecological security in the upper reaches of Yellow River has been affected significantly by land desertification in Gonghe Basin, especially Longyangxia Reservoir [17]. In order to control and prevent land desertification, large areas were planted with trees and shrubs to form the shelterbelt system in the ecotone between oasis and desert to protect farms, villages and roads in Gonghe Basin. Some trees in *Populus* were planted by transplanting inside oasis. Some shrubs in *Caragana* were planted by seeding on sand dunes. Other shrubs in *Salix* were planted by cutting on interdune.

In this chapter, the main water source of four dominant desert shrubs is compared by stable hydrogen and oxygen isotopes in the growing season (May, July and September). The water use ratio of different sources was analysed by Iso-source 1.3.1 software [18]. Two shrubs in *Caragana* on sand dunes may use shallow soil water recharged by rain and deep soil water recharged by ground water. Two shrubs in *Salix* on interdune may use shallow soil water recharged by rain, deep soil water recharged by ground water and even ground water. Further, the long-term WUE is compared using their leaf stable carbon isotope. The WUE of four shrubs may be higher in drought spring than in summer and autumn.

2. Water source and long-term WUE of four desert shrubs

2.1. Four dominant desert shrubs in Gonghe Basin

This study was conducted at Gonghe Desert Ecosystem Research Station, which is located in the mid-west Gonghe Basin and belongs to Shazhuyu Town, Gonghe County (36°16'N, 100°16'E, altitude 2874 m). The mean annual air temperature is only 2.4 °C, and the mean annual forest free day is only 91 days. The mean annual precipitation is 246.3 mm, which is concentrated from July to September. Two shrubs in *Caragana* and *Salix psammophila* were introduced from Ordos City, Inner Mongolia Autonomous Region. *Salix cheilophila* was introduced from Guinan County, which is located in the eastern Gonghe Basin.

Caragana intermedia (pea shrub) is a shrub with a height of 0.7–1.5 m growing on fixed sand dunes or flat sandy land. It is used as a good sand-fixing and afforest plant [19]. The mean height of *C. intermedia* was 1.04 m at Gonghe Station. The root system of *C. intermedia* was as deep as 1.3 m at Gonghe Station [20]. *Caragana korshinskii* (white pea shrub) is a shrub or small tree with a height of 1–4 m growing on fixed sand dunes. It is a good sand-fixing and soil conservation plant [21]. The mean height of *C. korshinskii* was 1.82 m at Gonghe Station.

S. psammophila (sandy willow) is a shrub with a height of 2–4 m, which inhabits on moving or semi-fixed sand dunes and interdune. *S. psammophila* is fast growing, tolerant to drought and sand burial and a good sand-fixing plant [21]. The mean height of *S. psammophila* was 2.82 m at Gonghe Station. The fine root of *S. psammophila* was 1.5 m deep in its original distribution area (Mu Us Sandy Land) [22]. *S. cheilophila* (black willow) is a small tree or big shrub, which is as tall as 5.4 m and inhabits slope, valley and riverbank. *S. cheilophila* is a mesophyte and hygrophyte and is often used as a sand-fixing or riparian plant [23]. The mean height of *S. cheilophila* was 3.72 m at Gonghe Station. The root system of *S. cheilophila* was 2.0 m deep at Gonghe Station [24].

2.2. Precipitation in the growing season and soil moisture in four desert shrubs plantation

The total precipitation at the study site was 137.5 mm from May 1 to September 13, 2014. The monthly precipitation was 8.8, 61.6, 37.8 and 28.7 mm in May, June, July and August, respectively (**Figure 1**). The maximal daily precipitation (18.9 mm) occurred on June 12. Before field sampling in July, 6.4 mm rain occurred on July 8, and in September, the precipitation was only 0.4 mm.

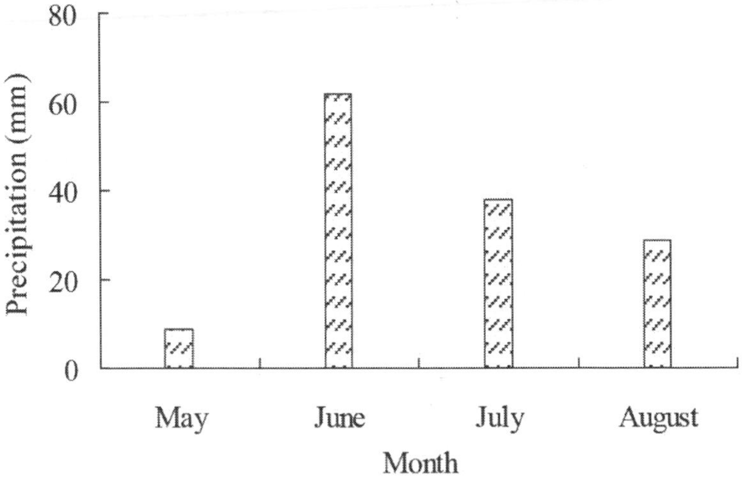

Figure 1. Precipitation at Gonghe Station from May to August, 2014.

Soil water content in *C. intermedia* plantation was significantly different in different depths on May 27 and July 20 but was similar on September 11 (**Figure 2**). On May 27, the water content in shallow soil (10 and 25 cm) was significantly higher than that in deeper soil (75–150 cm). On July 20, the water content in surface soil (10 cm) and subsurface soil (25 cm) was significantly higher than that in deeper soil (50–150 cm).

Soil water content in *C. korshinskii* plantation was similar in different depths on May 27 but was significantly different in different depths on July 23 and September 11 (**Figure 2**). On July 23, the water content in surface soil (10 cm) was significantly higher than in deeper depths (25–150 cm), and the water content in 150 cm was significantly higher than in medium depths (25–100 cm). On September 11, the water content in 150 cm was significantly higher than in shallower soil (10–100 cm).

Soil water content in *S. psammophila* plantation was significantly different in different depths on May 26, July 17 and September 11 (**Figure 2**). On May 26, the water content in middle and deep soil (50–150 cm) was significantly higher than that in shallow soil (10 and 25 cm). On July 17, the water content in middle and deep soil (50–150 cm) was significantly higher than that in shallow soil (25 cm). On September 11, the water content in middle soil (50 cm) and deep soil (150 cm) was significantly higher than that in shallow soil (10 and 25 cm).

Soil water content in *S. cheilophila* plantation was significantly different in different depths on May 26 and September 11 but was similar in different depths on July 17 (**Figure 2**). On May 26, the water content in middle and deep soil (50–200 cm) was significantly higher than that in surface soil (10 cm). On September 11, the water content in deep soil (100–200 cm) was significantly higher than in surface soil (10 cm) and middle soil (50 cm).

http://dx.doi.org/10.5772/63195

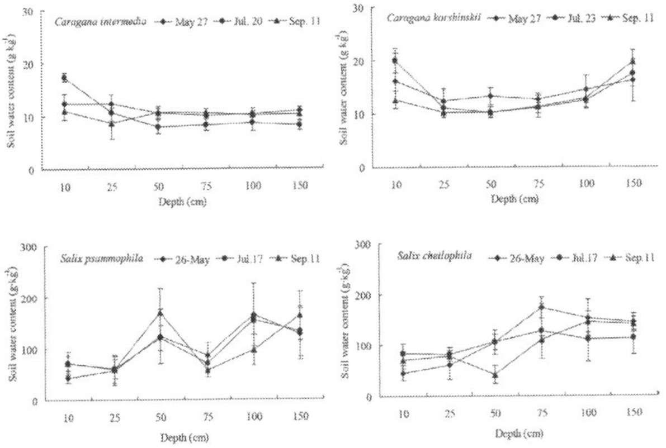

Figure 2. Soil water content in different depths in four desert shrub plantations.

2.3. Water source of four desert shrubs in different seasons

The $\delta^{18}O$ ratio of xylem water of *C. intermedia, C. korshinskii, S. psammophila* and *S. cheilophila* was located in the right side of global meteoric water line (**Figure 3**), which indicated that the water source of four shrubs was affected by isotope enrichment induced by evaporation. The δD and $\delta^{18}O$ value of some soil water were closer to ground water, which indicated deep soil water was recharged by ground water. The δD and $\delta^{18}O$ value of some soil water were closer to some rainwater, which indicated that shallow soil water is recharged by rain. The δD and $\delta^{18}O$ value of xylem water of two *Salix* shrubs were closer to ground water, which indicated that they might use ground water.

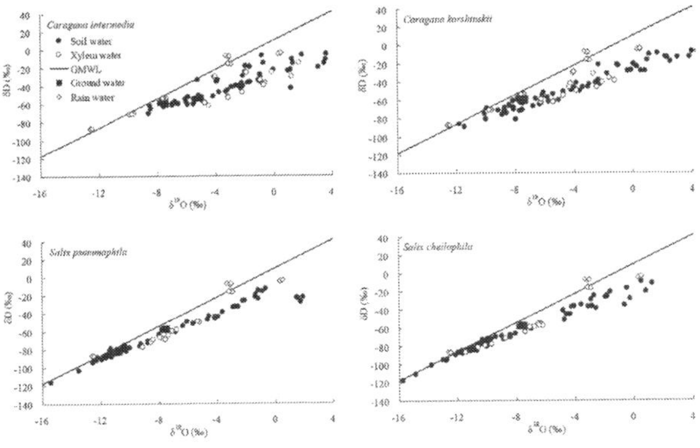

Figure 3. The value of δD and $\delta^{18}O$ in xylem and soil water of four desert shrubs, ground water, rainwater and global meteoric water line (GMWL) [4].

2.3.1. Water source of two Caragana shrubs on sand dunes

On May 27, the value of δD and $\delta^{18}O$ in xylem water of *C. intermedia* was closer to soil water in 50 cm (**Figure 4**). On July 20, the value of δD and $\delta^{18}O$ in xylem water of *C. intermedia* was closer to soil water in 10–50 cm, and the value of δD and $\delta^{18}O$ of soil water in 10 cm was closer to rain water on July 8 (6.4 mm). On September 11, the value of δD and $\delta^{18}O$ in xylem water of *C. intermedia* was closer to soil water in 100–150 cm and ground water.

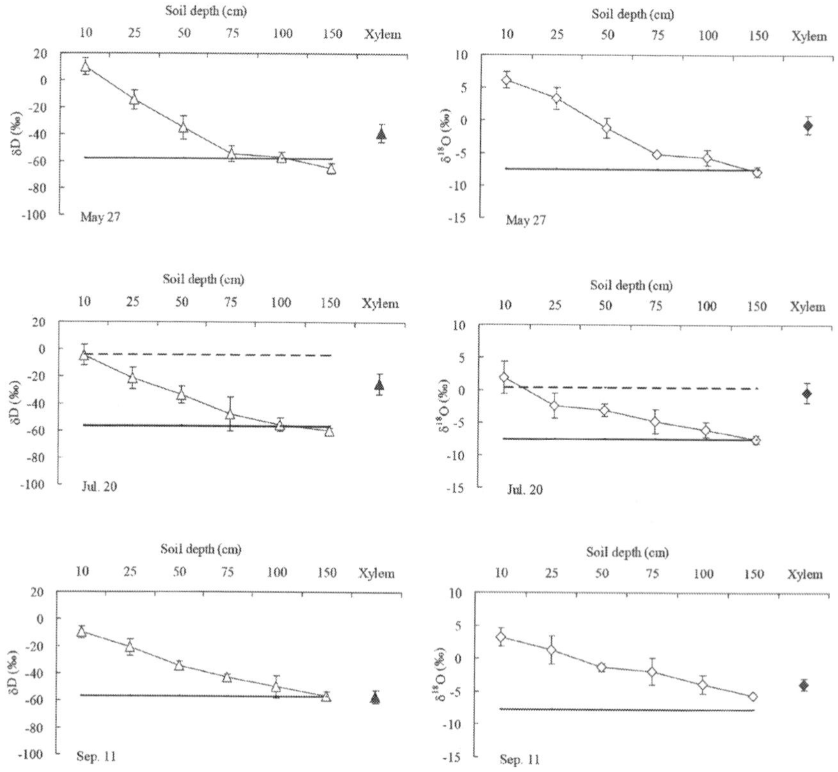

Figure 4. The value of δD and $\delta^{18}O$ in xylem water of *Caragana intermedia*, soil water and ground water. Full line is ground water. Dot line is 6.4 mm rainwater on July 8.

On May 27, the ratio of δD and $\delta^{18}O$ in xylem water of *C. korshinskii* was closer to soil water in 50–100 cm and ground water (**Figure 5**). On July 23, the value of δD and $\delta^{18}O$ in xylem water of *C. korshinskii* was closer to soil water in 50–75 cm, which was closer to rain water on July 21 (8.0 mm). On September 11, the value of δD and $\delta^{18}O$ in xylem water of *C. korshinskii* was closer to soil water in 25–75 cm.

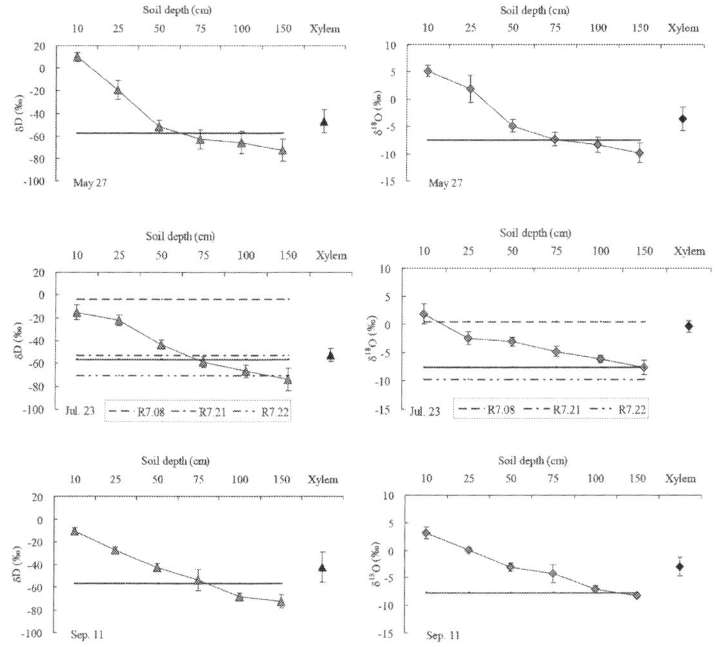

Figure 5. The value of δD and $\delta^{18}O$ in xylem water of *Caragana korshinskii*, soil water and ground water. Full line is ground water. R is rainwater.

Iso-Source analysis showed that *C. intermedia* mainly used 10–100 cm soil water on May 27, which accounted for 82.0% of its total water source. On July 20, *C. intermedia* mainly used 10 cm soil water, which accounted for 65.7% of its total water source. On September 11, *C. intermedia* mainly used 50–150 cm soil water and ground water, which accounted for 86.5% of its total water source (**Table 1**).

Water source	May 27	July 20	September 11
Soil water 10 cm	22.8 ± 12.2	65.7 ± 4.5	6.1 ± 5.3
25 cm	22.0 ± 15.3	8.8 ± 7.9	7.4 ± 6.5
50 cm	15.9 ± 13.8	7.6 ± 6.9	10.5 ± 9.1
75 cm	10.9 ± 9.3	5.8 ± 5.3	11.6 ± 1.0
100 cm	10.4 ± 8.9	4.5 ± 4.2	16.7 ± 14.5
150 cm	8.9 ± 7.5	3.7 ± 3.6	22.2 ± 16.8
Ground water	9.1 ± 7.8	3.8 ± 3.6	25.5 ± 12.9

Table 1. Water use ratio of *Caragana intermedia* to different sources (%, mean ± SD).

Iso-Source analysis showed that *C. korshinskii* evenly used 10–150 cm soil water and ground water on May 27. On July 23, *C. korshinskii* mainly used 25–150 cm soil water, which accounted for 82.8% of its total water source. On September 11, *C. korshinskii* mainly used 10–150 cm soil water, which accounted for 94.6% of its total water source (**Table 2**).

Water source	May 27	July 23	September 11
Soil water 10 cm	16.8 ± 9.3	9.7 ± 7.3	18.2 ± 10.6
25 cm	18.3 ± 12.2	12.0 ± 8.1	20.9 ± 14.5
50 cm	15.5 ± 13.4	17.9 ± 13.9	20.0 ± 16.6
75 cm	13.1 ± 11.1	13.9 ± 10.8	11.1 ± 8.2
100 cm	12.3 ± 10.4	10.4 ± 7.6	12.8 ± 9.2
150 cm	11.2 ± 9.4	28.6 ± 6.1	11.6 ± 7.8
Ground water	12.9 ± 11.0	7.5 ± 5.8	5.4 ± 3.7

Table 2. Water use ratio of *Caragana korshinskii* to different sources (%, mean ± SD).

On sand dunes of Gonghe Basin, *C. intermedia* used 10–100 cm soil water in spring, 10 cm shallow soil water recharged by rain in summer and 50–150 cm medium and deep soil water and ground water in autumn (**Figure 4, Table 1**). However, *C. korshinskii* used 10–150 cm soil water and ground water in spring, 25–150 cm soil water in summer and 10–150 cm soil water in autumn (**Figure 5, Table 2**). Therefore, two *Caragana* shrubs used shallow soil water recharged by rain and deep soil water recharged by ground water in spring or autumn, and even used few ground water when soil water is not enough for their water requirements (**Figure 2**). In summer, surface soil water recharged by rain was enough for *C. intermedia* (**Figures 1** and **4**), whereas *C. korshinskii* still needed both of shallow water recharged by rain and deep soil water recharged by ground water (**Figures 1** and **5**).

The resource-dependent water use strategy of two *Caragana* shrubs on sand dunes is an ecological adaptation to the semi-arid climate in Gonghe Basin, which is similar to other shrubs in arid and semi-arid regions. *A. ordosica* mainly used soil water within 50 cm in Mu Us Sandy Land [12]. In San Luis valley of Colorado, *Chrysothamnus greenei* only occurs in sites with water table deeper than 2 m and only use soil water recharged by rain [25]. *Caragana microphylla* mainly used 100 cm deep soil water recharged by snow in Xilin River Basin of Inner Mongolia [26]. Both of *N. tangutorum* and *A. arenaria* used 30–90 cm middle soil water in May, within 30 cm shallow soil water in July and below 120 cm deep soil water in September on sand dunes of Badain Jaran Desert [9]. *Ceratoides lateens* used 10–50 cm soil water in Qaidam Basin [27]. In Gurbantunggut Desert, *H. persicum* used 40–100 cm middle soil water on sand dunes when upper soil was abundant in early spring; however, it used 100–300 cm deep soil water when upper soil was depleted in summer [11].

The temporal difference in the main water source for a typical desert shrub is associated with precipitation change in different years. In this study, the ground water table is less than 3 m on interdune. *C. intermedia* also used ground water in September when precipitation was only 137.5 mm in the total growing season. However, *C. intermedia* mainly used soil water within 50 cm recharged by rain in middle August of 2009 with 155.5 mm precipitation since May [20]. On the other side, the spatial difference in the main water source for a typical desert shrub is associated with ground water table in different sites. *C. korshinskii* used 0–30 cm and 60–90 cm soil water in Ulanbuh Desert with ground water table deeper than 6 m [28]. At sites with water table less than 2 m in San Luis Valley, *S. vermiculatus* and *Chrysothamnus nauseosus* used top 50 cm soil water recharged by rain in a wet year. At sites with water table deeper than 2 m, they used ground water in the pre-monsoon season, whereas they used 30–40 cm upper soil water recharged by rain in the monsoon season [25].

2.3.2. *Water source of two Salix shrubs on interdune*

On May 26 and July 17, the value of δD and $\delta^{18}O$ in xylem water of *S. psammophila* was closer to ground water and soil water at depths of 25 and 50 cm (**Figure 6**). On July 17, the value of δD and $\delta^{18}O$ of soil water at a depth of 10 cm was closer to rainwater on July 8 (6.4 mm). On September 11, the value of δD and $\delta^{18}O$ in xylem water of *S. psammophila* was closer to that of ground water and soil water at depths of 50–150 cm.

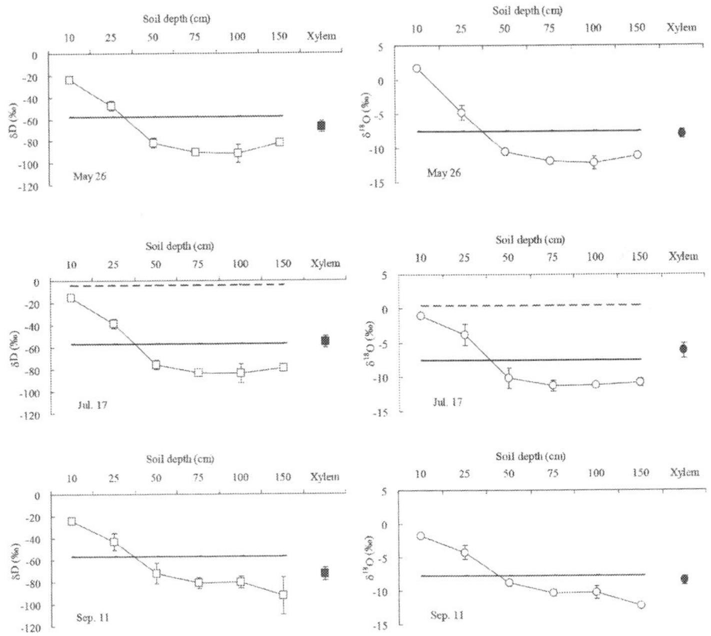

Figure 6. The value of δD and $\delta^{18}O$ in xylem water of *Salix psammophila*, soil water and ground water. Full line is ground water. Dot line is 6.4 mm rainwater on July 8.

On May 26, the value of δD and $\delta^{18}O$ in xylem water of *S. cheilophila* was closer to ground water and soil water in 25 and 50 cm (**Figure 7**). On July 17, the value of δD and $\delta^{18}O$ in xylem water of *S. cheilophila* was closer to ground water; and the value of δD and $\delta^{18}O$ of soil water in 10 cm was closer to rainwater on July 8 (6.4 mm). On September 11, the value of δD and $\delta^{18}O$ in xylem water of *S. cheilophila* was closer to ground water and soil water in 50 and 100–200 cm.

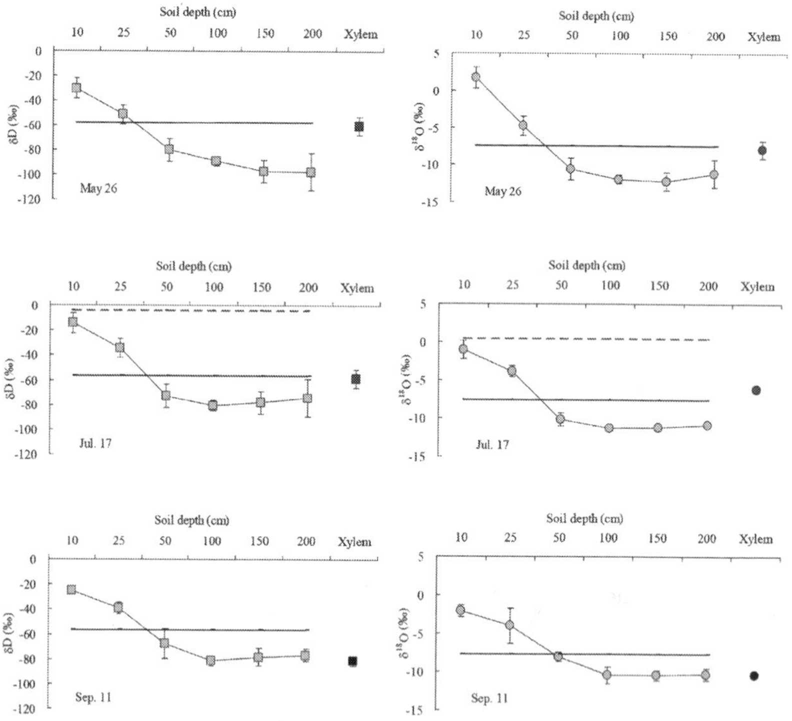

Figure 7. The value of δD and $\delta^{18}O$ in xylem water of *Salix cheilophila*, soil water and ground water. Full line is ground water. Dot line is 6.4 mm rainwater on July 8.

Iso-Source analysis showed that *S. psammophila* evenly used 10–150 cm soil water and ground water on May 26. On July 17, *S. psammophila* mainly used 10 and 25 cm soil water and ground water, which accounted for 63.0% of its total water source. On September 11, *S. psammophila* mainly used 50–150 cm soil water and ground water, which accounted for 79.6% of its total water source (**Table 3**).

Iso-Source analysis showed that *S. cheilophila* mainly used 10 and 25 cm soil water and ground water on May 26, which accounted for 70.7% of its total water source. On July 17, *S. cheilophila* evenly used 10–200 cm soil water and ground water. On September 11, *S. cheilophila* mainly used 100–200 cm soil water, which accounted for 81.8% of its total water source (**Table 4**).

http://dx.doi.org/10.5772/63195

Water source	May 26	July 17	September 11
Soil water 10 cm	13.0 ± 8.0	26.2 ± 12.0	9.0 ± 6.3
25 cm	16.0 ± 12.0	23.0 ± 16.9	11.3 ± 8.4
50 cm	14.3 ± 11.9	9.9 ± 8.6	16.8 ± 14.6
75 cm	12.8 ± 10.4	8.9 ± 7.7	16.1 ± 13.1
100 cm	12.6 ± 10.2	8.9 ± 7.7	16.2 ± 13.2
150 cm	14.3 ± 11.9	9.3 ± 8.0	14.4 ± 10.9
Ground water	17.0 ± 14.1	13.8 ± 12.0	16.1 ± 13.7

Table 3. Water use ratio of *Salix psammophila* to different sources (%, mean ± SD).

Water source	May 26	July 17	September 11
Soil water 10 cm	16.7 ± 7.1	17.0 ± 8.8	2.2 ± 2.1
25 cm	31.4 ± 20.2	18.8 ± 13.2	2.2 ± 2.1
50 cm	9.2 ± 7.7	13.1 ± 11.2	7.6 ± 6.8
100 cm	7.7 ± 6.4	11.5 ± 9.7	27.9 ± 19.9
150 cm	6.2 ± 5.2	11.7 ± 9.9	28.0 ± 19.9
200 cm	6.2 ± 5.2	12.3 ± 10.5	25.9 ± 19.9
Ground water	22.6 ± 12.2	15.6 ± 13.6	5.5 ± 5.0

Table 4. Water use ratio of *Salix cheilophila* to different sources (%, mean ± SD).

On interdune of Gonghe Basin, *S. psammophila* used 10–150 cm soil water and ground water in spring, 10–25 cm shallow soil water and ground water in summer and 50–150 cm medium and deep soil water and ground water in autumn (**Figure 6**, **Table 3**). However, *S. cheilophila* used 10–25 cm shallow soil water and ground water in spring, 10–200 cm soil water and ground water in summer and 100–200 cm deep soil water in autumn (**Figure 7**, **Table 4**). Therefore, two *Salix* shrubs used ground water or deep soil water recharged by ground water as long-term stable water source. Moreover, *S. psammophila* and *S. cheilophila* also used shallow soil water recharged by rain in spring and summer, respectively (**Figures 1**, **6**, and **7**).

The resource-dependent water use strategy of two *Salix* shrubs in Gonghe Basin is similar to other trees or shrubs in arid and semi-arid regions. *S. matsudana* and *S. vulgaris* mainly used deep soil water and ground water in Mu Us Sandy Land [12]. In San Luis Valley of Colorado, *S. vermiculatus* and *C. nauseosus* used soil water recharged by rain in a wet year but deep soil water and ground water in a dry year [25]. *E. nauseosa* depended on ground water, whereas *S. vermiculatus* used ground water during dry periods but used deep soil water after large rainfall in San Luis valley [10]. *J. osteosperma* absorbed shallow soil water in early spring and gradually depended on deep soil water with soil drought in Utah [6]. *P. sylvestris* var. *mongolica* primarily used 20–60 cm soil water both at the top of fixed sand dune and in interdune lowland in Horqin Sandy Land [29]. Moreover, it only used soil water during the higher precipitation year, whereas it used soil water and relied on ground water during the lower precipitation year [15]. *U. pumila* always used stable ground water in Hunshandake Sandy Land [14]. *N. tangutorum* mainly used 50–100 cm soil water and ground water from June to September in Golmud of Qaidam Basin [30]. In Nuomuhong of Qaidam Basin, *N. tangutorum* and *Tamarix ramosissima* used 50–70 cm soil water, whereas *Ephedra sinica* and *Calligonum mongolicum* evenly used 0–90 cm soil water. Moreover, these four shrubs increased to use ground water in the late growing season [27]. Along the lower reaches of Heihe River, different aged *P. euphratica* used different water sources. Young forests primarily relied on soil water from 0 to 50 cm (mean > 45%), while mature and over-mature forests used water from deeper than 100 cm which derived primarily from ground water [16]. When the upper soil water was abundant in early spring, *H. ammodendron* mainly used shallow soil water on sand dunes of Gurbantunggut Desert. When the upper soil water was depleted in summer, *H. ammodendron* mainly used ground water [11]. Therefore, ground water is an important water source for trees and shrubs in arid and semi-arid regions, especially in the drought period. Global climate change may result in the increase of extreme drought in the arid region [31]. Plant species using stable ground water or deep soil water may have more adaptability than those only using shallow soil water recharged by rain in the drought period.

2.4. Long-term WUE of four desert shrubs

Leaf carbon discrimination (Δ, ‰) was converted from its $\delta^{13}C$ value using an atmospheric carbon dioxide ratio of −8‰ [32]. The leaf Δ value was significantly different ($P < 0.001$) in different months for *C. intermedia, C. korshinskii, S. psammophila* and *S. cheilophila*, respectively (**Figure 8**). The leaf Δ value of *C. intermedia, C. korshinskii* and *S. psammophila* increased significantly on May 26, and from July 17 to September 11. The leaf Δ value of *S. cheilophila* was significantly lower on May 26 than that on July 17 or September 11. The leaf Δ value of *C. intermedia* was similar to *C. korshinskii* in three months. The leaf Δ value of *S. psammophila* was lower than *S. cheilophila* in three months.

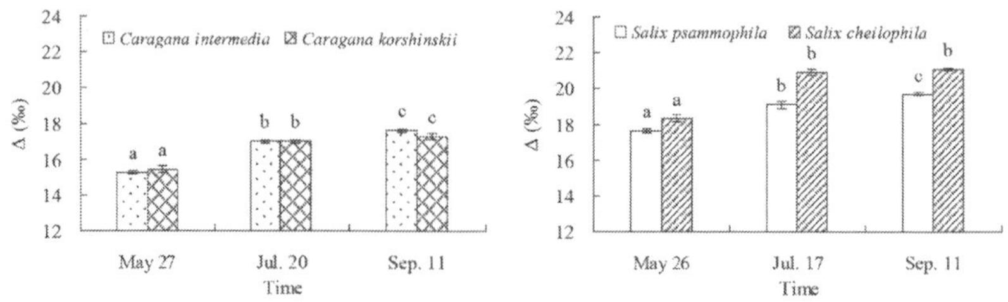

Figure 8. Leaf carbon discrimination of four desert shrubs. Different lower case letters indicate significant difference in different months, according to Duncan's test ($P < 0.05$).

The long-term WUE is negatively related to the leaf Δ value of C_3 plants. There were seasonal dynamics of WUE in two *Caragana* shrubs on sand dunes of Gonghe Basin, which were lower in spring and increased in summer and autumn (**Figure 6**). Soil water content at surface soil was lower in spring than in summer (**Figure 2**) because of low precipitation (**Figure 1**). It was suggested that both of two *Caragana* shrubs increased WUE to adapt to drought. Similarly, the WUE of *N. tangutorum* was higher in summer than in autumn in Ulanbuh Desert [28]. The WUE of *P. euphratica* was highest in May and decreased in the middle and late growing season in lower reaches of Heihe River, west Inner Mongolia [33].

There were seasonal dynamics of long-term WUE indicated by the leaf Δ value in two *Salix* shrubs on interdune of Gonghe Basin, which were higher in spring than in summer and autumn (**Figure 7**). Soil water content of two *Salix* plantations at surface soil was lower than at deeper depths (**Figure 2**) because of low precipitation in spring (**Figure 1**). It was suggested that both of two *Salix* shrubs increased WUE to adapt to drought. There was intra-species difference in WUE between the two *Salix* shrubs. The WUE of *S. psammophila* was always higher than that of *S. cheilophila* (**Figure 8**), which is related to their eco-physiology. The xerophyte *S. psammophila* has higher drought resistance than the mesophyte *S. cheilophila*. Further, the WUE of different shrubs is affected by their life form. The evergreen *S. vulgaris* had higher WUE than *S. matsudana* and *A. ordosica* in Mu Us Sandy Land [12]. The long-term WUE of evergreen *Ammopiptanthus mongolicus, N. tangutorum* and *C. korshinskii* was higher than that of *A. ordosica* in Ulanbuh Desert [28]. In arid and semi-arid regions, shrubs with higher WUE may have more adaptability to extreme drought, resulting from global climate change. Moreover, the leaf $\delta^{13}C$ value (−25.11‰ – −27.28‰) of *S. psammophila* in Gonghe Basin is similar to its original distribution area (−26.75‰ – −28.68‰, Wushen Banner, Ordos City) [34], which means that this shrub has adapted to the semi-arid climate in Gonghe Basin.

2.5. Conclusion

The water use strategy of four dominant desert shrubs was adapted to the semi-arid climate in Gonghe Basin. They used different water sources depending on their availability in different seasons, including shallow soil water recharged by rain, deep soil water recharged by ground water or ground water. They could use shallow soil water after rain in spring and summer. When shallow soil water was depleted, they turned to use deep soil water or ground water. The reliance of ground water was different for four shrubs in two habitats. Two shrubs in *Caragana* mainly depend on soil water and did not depend on ground water on sand dunes. *C. intermedia* barely used ground water, whereas *C. korshinskii* only used ground water in spring, which accounted for 12.9% of its total water source. Two shrubs in *Salix* used ground water as an important stable water source on interdune. *S. psammophila* always used ground water in the growing season, which accounted for 13.1–16.1% of its total water source; whereas *S. cheilophila* used ground water in spring and summer, which accounted for 22.6% and 15.6% of its total water source. Moreover, both of four shrubs increased long-term WUE in drought spring. Two shrubs in *Caragana* had similar WUE in the growing season. However, the WUE of *S. psammophila* was always higher than *S. cheilophila*, which means that the former might had more adaptability in Gonghe Basin.

Acknowledgements

This study was supported by Chinese National Natural Science Foundation (41301095) and the Fundamental Research Funds for the Central Non-profit Research Institute of Chinese Academy of Forestry (CAFYBB2014QB023).

Author details

Yajuan Zhu*

Address all correspondence to: zhuyj@caf.ac.cn

Institute of Desertification Studies, Chinese Academy of Forestry, Beijing, China

References

[1] Ehleringer JR. Carbon and water relations in desert plants: an isotopic perspective. In: Ehleringer JR, Hall AE, Farquhar GD, editors. Stable Isotope and Plant Carbon–Water Relations. 1st ed. San Diego: Academic Press; 1993. p. 155–172. DOI: 10.1016/B978-0-08-091801-3.50018-0

[2] Schwinning S, Starr BI, Ehleringer JR. Dominant cold desert plants do not partition warm season precipitation by event size. Oecologia. 2003;136:252–260. DOI: 10.1007/s00442-003-1255-y

[3] Ogle K, Reynolds JF. Plant responses to precipitation in desert ecosystems: integrating functional types, pulses, thresholds, and delays. Oecologia. 2004;141:282–294. DOI: 10.1007/s00442-004-1507-5

[4] Lin G. Stable Isotope Ecology. 1st ed. Beijing: Higher Education Press; 2013. 492 p.

[5] Cheng X, An S, Li B, Chen J, Lin G, Liu Y, Luo Y, Liu S. Summer rain pulse size and rainwater uptake by three dominant desert plants in a desertified grassland ecosystem in northwestern China. Plant Ecology. 2006;184:1–12. DOI: 10.1007/s11258-005-9047-6

[6] West AG, Hultine KR, Burtch KG, Ehleringer JR. Seasonal variation in moisture use in a piňon-juniper woodland. Oecologia. 2007;153:787–798. DOI: 10.1007/s00442-007-0777-0

[7] Li S-G, Romero-Saltos H, Tsujimura M, Sugimoto A, Sasaki L, Davaa G, Oyunbaatar D. Plant water sources in the cold semiarid ecosystem of the upper Kherlen River catchment in Mongolia: a stable isotope approach. Journal of Hydrology. 2007;333:109–117. DOI: 10.1016/j.jhydrol.2006.07.020

[8] Kowaljow E, Fernández RJ. Different utilization of a shallow-water pulse by six shrub species in the Patagonian steppe. Journal of Arid Environments. 2011;75:211–214. DOI: 10.1016/j.jaridenv.2010.10.004

[9] Zhu Y, Jia Z, Yang X. Resource-dependent water use strategy of two desert shrubs on interdune, Northwest China. Journal of Food, Agriculture and Environment. 2011;9:832–835.

[10] Kray JA, Cooper DJ, Sanderson JS. Groundwater use by native plants in response to changes in precipitation in an intermountain basin. Journal of Arid Environments. 2012;83:25–34. DOI: 10.1016/j.jaridenv.2012.03.009

[11] Dai Y, Zheng X-J, Tang L-S, Li Y. Stable oxygen isotopes reveal distinct water use patterns of two *Haloxylon* species in the Gurbantonggut Desert. Plant and Soil. 2015;389:73–87. DOI: 10.1007/s11104-014-2342-z

[12] Ohte N, Koba K, Yoshikawa K, Sugimoto A, Matsuo N, Kabeya N, Wang L. Water utilization of trees in semiarid desert of Inner Mongolia, China. Ecological Applications. 2003;13:337–351. DOI: 10.1890/1051-0761

[13] Pataki DE, Bush SE, Gardner P, Solomon DK, Ehleringer JR. Ecohydrology in a Colorado River riparian forest: implications for the decline of *Populus fremontii*. Ecological Applications. 2005;15:1009–1018. DOI: 10.1890/04-1272

[14] Su H, Li Y, Liu W, Xu H, Sun OJ. Changes in water use with growth in *Ulmus pumila* in semiarid sandy land of northern China. Trees: Structure and Function. 2014;28:41–52. DOI: 10.1007/s00468-013-0928-3

[15] Song L, Zhu J, Li M, Yu Z. Water utilization of *Pinus sylvestris* var. *mongolica* in a sparse wood grassland in the semiarid sandy region of Northeast China. Trees: Structure and Function. 2014;28:971–982. DOI: 10.1007/s00468-014-1010-5

[16] Liu S, Chen Y, Chen Y, Chen Y, Friedman JM, Hati JHA, Fang G. Use of ^{2}H and ^{18}O stable isotopes to investigate water sources for different ages of *Populus euphratica* along the lower Heihe River. Ecological Restoration. 2015;30:581–587. DOI: 10.1007/s11284-015-1270-6

[17] Zhang D, Gao S, Shi M, Hasi E, Yan P, Lu R. Land Desertification and Its Control in Plateau of Qinghai. 1st ed. Beijing: Science Press; 2009. 194 p.

[18] Phillips DL, Gregg JW. Source partitioning using stable isotopes: coping with too many sources. Oecologia. 2003;136:261–269. DOI: 10.1007/s00442-003-1218-3

[19] Liu YX, editor. Flora in Desertis Reipublicae Populorum Sinarum. Tomus 2. 1st ed. Beijing: Science Press; 1987. 464 p.

[20] Jia Z, Zhu Y, Liu L. Different water use strategies of juvenile and adult *Caragana intermedia* plantations in the Gonghe Basin, Tibet Plateau. PLoS One. 2012;7:e45902. DOI: 10.1371/journal.pone.0045902

[21] Liu SW, editor. Flora Qinghaiica. Volume 1. 1st ed. Xining: Qinghai Pepole's Publishing House; 1997. 544 p.

[22] Liu J, He X, Bao H, Zhou C. Distribution of fine roots of *Salix psammophila* and its relationship with soil moisture in Mu Us Sandland. Journal of Desert Research. 2010;30: 1362-1366.

[23] Ma YQ, editor. Flora Intramongolica. Volume 1. 1st ed. Huhhot: Inner Mongolia People Press; 1985. 294 p.

[24] Yu Y, Jia Z, Zhu Y, Zhao S, Liu H, Li Q, Wei D. Root distribution of *Salix cheilophila* along a chronosequence in high-cold sandland. Journal of Desert Research. 2014;34:67–74. DOI: 10.7522/j.issn.1000-694X.2013.00188

[25] Chimner RA, Cooper DJ. Using stable oxygen isotopes to qualify the water source used for transpiration by native shrubs in San Luis Valley, Colorado USA. Plant and Soil. 2004;260:225–236. DOI: 10.1023/B:PLSO.0000030190.70085.e9

[26] Yang H, Auerswald K, Bai Y, Han X. Complementarity in water sources among dominant species in typical steppe ecosystems of Inner Mongolia, China. Plant and Soil. 2011;340:303–313. DOI: 10.1007/s11104-010-0307-4

[27] Xing X, Chen H, Zhu J, Chen T. Water sources of five dominant desert species in Nuomuhong area of Qaidam Basin. Acta Ecologica Sinica. 2014;34:6277–6286. DOI: 10.5846/stxb201310092427

[28] Zhu Y, Jia Z, Lu Q, Hao Y, Zhang J, Li L, Qi Y. Water use strategy of five shrubs in Ulanbuh Desert. Scientia Silvae Sinicae. 2010;46:15–21. DOI: 10.11707/j.1001-7488.20100403

[29] Wei YF, Fang J, Liu S, Zhao XY, Li SG. Stable isotope observation of water use source of *Pinus sylvestris* var. *mongolica* in Horqin Sandy Land, China. Trees: Structure and Function. 2013;27:1249–1260. DOI: 10.1007/s00468-013-0873-1

[30] Gong G, Chen H, Duan D. Comparison of the methods using stable hydrogen and oxygen isotope to distinguish the water source of *Nitraria tangutorum*. Acta Ecologica Sinica. 2011;31:7533-7541

[31] IPCC, editor. Climate Change 2013. The Physical Science Basis. New York: Cambridge University Press; 2013. 152 p.

[32] West AG, Bowen GJ, Cerling TE, Ehleringer JR. Stable isotopes as one of nature's ecological recorders. TRENDS in Ecology and Evolution. 2006;21:408-414. DOI: 10.1016/j.tree.2006.04.002

[33] Cao S, Feng Q, Si J, Chang Z. Research on characteristics of water use efficiency variations of *Populus euphratica* during the growing season in extremely arid region of China. Journal of Desert Research. 2012;32:724–729.

http://dx.doi.org/10.5772/63195

[34] Liu H, Li J, Zhao Y, Huang K. Study on $\delta^{13}C$ values of desert plant *Salix psammopphila* and water use efficiency. Arid Zone Research. 2008;25:514–518.

Chapter 6

Intelligent Soft Computing Models in Water Demand Forecasting

Sina Shabani, Peyman Yousefi, Jan Adamowski and
Gholamreza Naser

Additional information is available at the end of the chapter

http://dx.doi.org/10.5772/63675

Abstract

Given the increasing trend in water scarcity, which threatens a number of regions worldwide, governments and water distribution system (WDS) operators have sought accurate methods of estimating water demands. While investigators have proposed stochastic and deterministic techniques to model water demands in urban WDS, the performance of soft computing techniques [*e.g.*, Genetic Expression Programming (GEP)] and machine learning methods [*e.g.*, Support Vector Machines (SVM)] in this endeavour remains to be evaluated. The present study proposed a new rationale and a novel technique in forecasting water demand. Phase space reconstruction was used to feed the determinants of water demand with proper lag times, followed by development of GEP and SVM models. The relative accuracy of the three best models was evaluated on the basis of performance indices: coefficient of determination (R^2), mean absolute error (*MAE*), root mean square of error (*RMSE*), and Nash-Sutcliff coefficient (*E*). Results showed GEP models were highly sensitive to data classification, genetic operators, and optimum lag time. The SVM model that implemented a Polynomial kernel function slightly outperformed the GEP models. This study showed how phase space reconstruction could potentially improve water demand forecasts using soft computing techniques.

Keywords: water demand forecasting, soft computing, genetic expression programming, support vector machines, phase space reconstruction, lag time

INTECH
open science | open minds

© 2016 The Author(s). Licensee InTech. This chapter is distributed under the terms of the Creative Commons Attribution License (http://creativecommons.org/licenses/by/3.0), which permits unrestricted use, distribution, and reproduction in any medium, provided the original work is properly cited.

1. Introduction

While water scarcity has become a key concern worldwide, it is particularly so in arid and semiarid regions with limited potable water sources. In designing water distribution systems (WDS), engineers have typically used a "fixture unit" method, which considers the sum of fixture unit demands, facility types, and socioeconomic factors to determine peak demand. However, this overestimates the actual peak demand by as much as 100% [1]. Due to various uncertainties, including those associated with demand, engineers often include large safety factors when designing WDS. Given that WDS rely mainly on regional energy and resources, an overdesigned system can have environmental impacts that will appear in region(s) beyond the jurisdictional boundaries of the system. While short-term demand forecasts are critical to a WDS daily operations [2], long-term forecasts are required for future planning and management of the systems. In providing an accurate estimate of water demand, a robust demand-forecasting model assists managers in designing a more environmentally sustainable WDS and in managing available water resources more efficiently. When coupled with a water demand management strategy, such models can help managers overcome operational problems (e.g., low pressure during peak demands) and issues related to asset management (e.g., nonreplacement of assets or replacement by lower capacity assets reaching the end of their economic life). It has been estimated that a well-predicted monthly average demand might be up to 400% lower than peak demands that cause low pressure; however, a more realistic model can enhance resource management and operating systems. This will eventually lead to significant savings for water and energy (for running pumps, treatment plants, etc.) industries. Considering weather conditions and population, the prime objective of the present study was to develop a predictive model for monthly average water demand. While the present study proposed a generic framework that could be easily adjusted for any specific case, the City of Kelowna (British Columbia, Canada) was employed as a test case.

2. Literature review

Water demand varies greatly both regionally and seasonally. Increasing urbanization and industrialization as well as emerging issues such as shifting weather patterns and population growth have significant impacts on water demand. The main components in demand prediction are the explanatory variables and time scales used. Selecting explanatory variables for a predictive model depend on the desired time scale and the availability of data. Simple models using very few explanatory variables have shown promising accuracy for short-term prediction [3, 4]. In general, the explanatory variables affecting water demand are of two types: weather (e.g., temperature, relative humidity, and rainfall) and socioeconomic (e.g., population and income). Weather conditions affect short-term prediction while their socioeconomic counterparts can affect long-term predictions [5–7]. As has been highlighted by significant worldwide changes in climate, both in terms of weather conditions and global warming, water availability is prone to great uncertainty [8]. Therefore, the impact of evolving weather conditions on long-term water demand predictions should receive greater attention. Furthermore, researchers who have considered weather conditions in short-term water demand prediction have established that it is not feasible to feed online automated WDS with real time weather information [9]. As a result, limited studies have considered weather conditions in their demand forecasting models [10–12]. **Table 1** summarizes the relevant literature. Temperature, precipitation, pan evaporation, and number of days since the last rainfall were used in a forecasting model [13]. Another study used temperature, relative humidity, rainfall, wind speed, and air pressure as weather parameters in their hourly water demand model for Sao Paulo, Brazil [12]. **Table 1** shows the previous researchers did not consider socioeconomic and weather conditions simultaneously since their effects are highly dependent on the forecast's time scale. Traditionally, WDS utilities have used historical patterns as explanatory variables in predicting future water demands. Scarce water reserves and the rapid increase in urbanization have raised awareness and led to implementation of statistical approaches. Multiple linear regression (MLR) and time series were the most popular techniques used in the early stages of demand forecasting [6]. While MLR has been widely used to better understand the determinants of water demand [14–18], its major drawback is the fact that it considers linear relationships among variables and water demand, such relationships are nonlinear by nature. Time series have been introduced along with regression as methods for demand forecasting [10, 19]. Due to the common belief that they can deal with complex systems [20], artificial neural networks (ANNs) have been widely applied in water demand forecasting [21–23, 2]. Comparing regression, univariate time series, and ANN models, one study found ANN models drawing on standard rainfall and maximum temperature data could better predict weekly water demand than other models [6]. Similarly, drawing on temperature and rainfall data in their forecasting models, researchers concluded that ANN models provided more reliable forecasts for peak weekly demand than time series and simple and multiple linear regressions [22]. Results of another study showed ANN models performed better for hourly forecasts, whereas regression models were more accurate in forecasting daily demand [23]. To improve the accuracy and robustness of demand forecasting models, hybrid models combining or modifying ANN, MLR, and time series techniques have been tested [24–27]. However, application of nonlinear regression in demand forecasting has remained limited to studies using support vector machines (SVMs) [28–30] and training nonlinear relationships through linear regression models [6, 31]. The present study compares gene expression programming (GEP) and SVM nonlinear approaches. Inspired by Darwin's theory of evolution, GEP was recently proposed in engineering disciplines to optimize the structure of input variables fed into predictive models [32]. Being a self-learning algorithm, GEP has several advantages over conventional predictive models. GEP defines individual block structures (input variables, response, and function sets) and selects the optimized operating functions and multipliers through the process of learning algorithms. Results of one study indicated GEP models outperformed traditional linear models in the field of hydrology [32]. Since weather information is one of the major determinants of water demand, this research employed GEP to develop a robust and accurate demand-forecasting model.

No.	Reference	Method	Determinant	Time scale
1	[16]	Linear regression	Seasonal dummies, derivatives of weather and price	Monthly demand
2	[17]	Linear regression	Density, building size, lot size, household size, income, price, temp, rain, drought dummies	Bimonthly demand
3	[18]	Regression using Bayesian moment entropy	Population density	Annual demand
4	[13]	Decomposed daily demand followed by composite forecasts	Daily demand and hourly demand	Daily demand
5	[19]	Univariate time series	Y_{t-1}	Annual residential demand
6	[22]	Regression and ANN	Temp, rainfall, and lags of peak demand	Peak weakly demand
7	[23]	ANN	Temperature, rainfall, and delayed demand	Daily demand
8	[2]	Time series	Univariate demand series, temperature in a multivariate model	Daily, weekly, monthly, annual
9	[6]	Time series and ANN	Delayed demands, temperature, and rainfall	Weekly demand
10	[24]	Holt-Winters multiplicative smoothing modified regression	Precipitation, temperature, humidity, lagged demand	Weekly (6 days)
11	[26]	Weighted average regression and ANN	Historical demand and time	Annual demand
12	[27]	Decomposed annual demand, regression and ANN	GDP, population, temperature, greenery coverage, delayed demand	Annual demand
13	[31]	Wavelet-deinoizing and ANN	7-year long time series of demand	Monthly demand
14	[28]	SVM with RBF function is compared with ANN	Delayed demand, population	Daily demand
15	[29]	ANN, SVM, Monte Carlo	Rain, demand, wind speed, atmospheric pressure	Hourly demand
16	[30]	SVM and adaptive Fourier series	Wind speed, temperature, demand, humidity, and rainfall	Hourly demand

Table 1. Literature on water demand forecasting.

3. Study area and data collection

This research focused on the City of Kelowna located in the Okanagan Valley (British Columbia, Canada). The City has five water districts including the City of Kelowna District (CKD), Glenmore Ellison Irrigation District (GEID), Black Mountain Irrigation District (BMID), Rutland Water District (RWD), and the South East Kelowna Irrigation District (SEKID). The CKD served as the study area of this research. Using three major pumping stations, the CKD primarily supplies water from the Okanagan Lake. The present study used monthly mean water demand data from 1996 to 2010 (http://www.kelowna.ca/). The population censuses of 1996, 2001, 2006, and 2011, along with the best-fit parabolic equation (with coefficient of determination of $R^2 \approx 1$) allowed estimation of the population in noncensus years. Weather indices including temperature, wind speed, relative humidity, and rainfall, were drawn from the Environment Canada weather data (http://kelowna.weatherstats.ca/) collected at Station A (latitude 49°57′13″N, longitude 119°22′29″W) located at the City of Kelowna's airport.

4. Methodology

4.1. Model development

To determine water demand (*D*) in millions of liters (ML), this research used population (*P*) and hotel occupancy factor (HOR) as socioeconomic parameters (the City of Kelowna is one of the hot spots for tourism in North America), and temperature (*T*) in °C, relative humidity (RH) in percent, and rainfall (*R*) in millimeters as weather parameters. As these parameters did not have the same order of magnitude, they were normalized prior to models development by

$$X = \frac{x - \mu}{\sigma} \tag{1}$$

where X is the standardized magnitude of parameter x, μ and σ are the corresponding mean and standard deviation, respectively. Phase space reconstruction of each explanatory variable was used prior to GEP modeling to define the structure of the model inputs. This was done to identify the stochastic or deterministic nature of the collected data. For a given proper lag time, the phase space was built by applying Taken's theorem [33] and transforming the time-series data into the geometry of a single moving point along a trajectory, where each point corresponds to a datum. Average mutual information (AMI) was used to determine the proper lag time of water demand for phase space reconstruction of all input factors. This was done to achieve a comprehensive understanding of input factors, variable self-interaction, and assess the use of lag times in demand forecasting models. Labeled $M_aD_bOP_c$, where a, b, and $c \in \{1, 2, 3\}$ a total of 27 models were created (**Table 2**), which combined three input types [M_1: demand data only; M_2: demand and climatic data; M_3: demand, climatic, and demographic data], three lag times [D_1: 1 month lag; D_2: 1 and 2 month lags; D_3: 1, 2, and 3 month lags], and three types of genetic operators [OP_1: {+, −, x}; OP_2: {+, −, x, x^2, x^3}; OP_3: {+, −, x, x^2, x^3, √, e^x, log, ln}] used in developing the GEP models.

Classification	Model	Input variables combination*
Demand Data Based	M_1D_1	D_{t-1}
	M_1D_2	D_{t-1}, D_{t-2}
	M_1D_3	D_{t-1}, D_{t-2}, D_{t-3}
Demand + Weather Data Based	M_2D_1	D_{t-1}, T_{t-1}, R_{t-1}, RH_{t-1}
	M_2D_2	D_{t-1}, D_{t-2}, T_{t-1}, T_{t-2}, R_{t-1}, R_{t-2}, RH_{t-1},
	M_2D_3	D_{t-1}, D_{t-2}, D_{t-3}, T_{t-1}, T_{t-2}, T_{t-3}, R_{t-1}, R_{t-2}, R_{t-3}, RH_{t-1}, RH_{t-2}, RH_{t-3}
Demand + Weather + Population Data Based	M_3D_1	D_{t-1}, T_{t-1}, R_{t-1}, RH_{t-1}, P, HOR
	M_3D_2	D_{t-1}, D_{t-2}, T_{t-1}, T_{t-2}, R_{t-1}, R_{t-2}, RH_{t-1}, RH_{t-2}, P, HOR
	M_3D_3	D_{t-1}, D_{t-2}, D_{t-3}, T_{t-1}, T_{t-2}, T_{t-3}, R_{t-1}, R_{t-2}, R_{t-3}, RH_{t-1}, RH_{t-2}, RH_{t-3}, P, HOR

*t is current month; D is demand; HOR is hotel occupancy factor; P, is population; R is rainfall; RH is relative humidity; T is temperature.

Table 2. Structure of classified models.

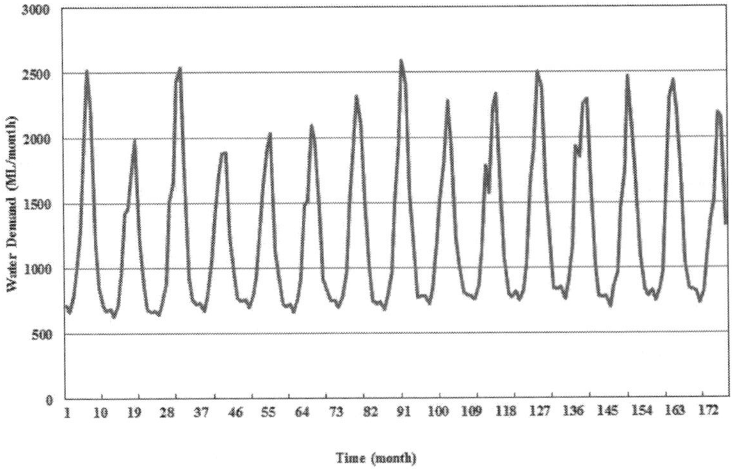

Figure 1. Time series of water demand in the City of Kelowna District (CKD) for 1966–2008.

Data were used in partitions of 144 samples for training (1996–2007) and 35 samples for validation (2008–2010). The time series of water demand over the time period of 1996–2010 (**Figure 1**) shows a relatively regular periodic cycle of water demand in CKD that is mainly due to seasonal changes.

4.2. Genetic expression programming (GEP)

Introduced by Ferreira, GEP is an emerging soft computing technique [34]. The strategy used for the learning algorithms was the optimal evolution using the genetic operators. Following Ferreira, this research defined the overall structure of the GEP model by: 30 chromosomes, eight head sizes, and three genes [35]. The selected head size determined how complex each model parameter was. Each of the gene heads underwent a set of different arrangements to model the feeding data. Selecting new random populations was followed by reproduction in order to reach the most suitable model under optimized stopping conditions. Models were developed based on three genes linked together by an addition function. The number of genes per chromosome specified the layers or blocks involved in building the whole model. Although a large gene was useful, dividing the chromosomes into simpler units resulted in a more efficient and manageable learning process. RMSE was used as a fitness function to fit a curve to target values. The stopping condition was a maximum fitness and coefficient of determination (R^2). Ten numerical constants were used as floating data point in each gene.

4.3. Lag time

The literature lists three methods for estimating lag time, AMI, autocorrelation function (ACF), and correlation integral (CI) [36–38]. AMI is considered the best since ACF reflects only linear properties and CI requires a large set of data [39]. Consequently, the present study employed AMI defined as:

$$I_\tau = \sum_{i=1}^{i=n} P(X_i, X_{i+\tau}) \cdot \log_2 \frac{P(X_i, X_{i+\tau})}{P(X_i) \cdot P(X_{i+\tau})} \tag{2}$$

where the joint probability of two successive time series, $P(X_i, X_{i+\tau})$ and the product of their individual marginal probability, $P(X_x) \cdot P(X_{i+x})$, were used to find the optimum lag time. This lag can contribute to the maximum information added on X_i by the successive time series $X_{i+\tau}$. The prime objective of using this approach was to make sure these time series were independent and thereby better represented the dynamics of the system in the phase space. In other words, a balanced independency was desirable in identifying an optimum delay time.

4.4. Support vector machines (SVM)

For SVM models, in which genetic operators are not used, the input types remained M_1, M_2, or M_3, while the lag times remained D_1, D_2, or D_3. This study compared the performance of radial basis function (RBF), polynomial (Poly), and Linear (Lin) kernels. These were appended to the input type and lag, e.g., M_1D_1RBF, M_1D_1Poly, *or* M_1D_1Lin. **Figure 2** shows the structure of the SVM model. Kernel functions (RBF, Poly, or Lin) were used to map the input vectors into higher dimensions in space.

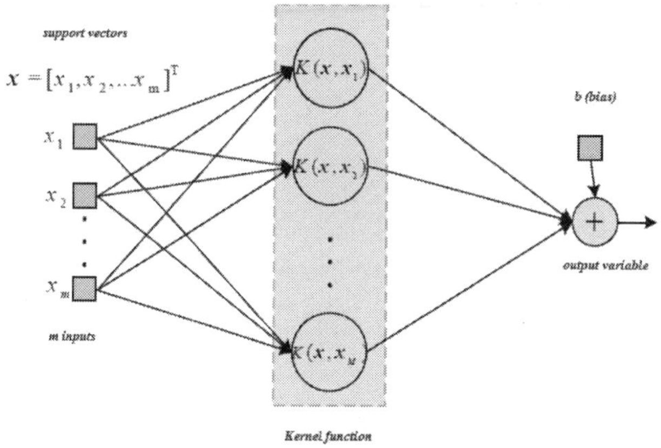

Figure 2. Support vector machine structure.

In this method, the input vectors are considered as supports forming the backbone of the whole model structure through a training process. If N samples of the population given by $X \in R^m$, $\{X_K, Y_K\}_{K=1}^{N}$, $Y \in R$, a function or SVM estimator on a regression can be considered as:

$$f(x) = W\phi(X) + b \tag{3}$$

where X is an input parameter with m components and Y is its response output variable, W is a weight vector, b represents a bias, and φ is a transfer function which exhibits nonlinear behavior, mapping the input vectors into a higher dimensional space. As these mapped vectors can compromise the complex nonlinear regression of the input space, Cortes and Vapnik introduced the convex optimization problem with an insensitivity loss function [40]:

$$\text{minimize}\, w, b, \xi, \xi^* \quad \frac{1}{2} \; W \;^2 + C \sum_{k=N}^{k=1} (\xi_k - \xi_k^*) \tag{4}$$

$$\text{subject to} \begin{Bmatrix} Y_k - W^T\phi(X_k) - b \le \varepsilon + \xi_k \\ W^T\phi(X_k) + b - Y_k \le \varepsilon + \xi_k^* \\ \xi_k, \xi_k^* \ge 0 \end{Bmatrix} k = 1, 2, \cdots, N \tag{5}$$

where ξ_k and ξ_k^* are slack variables that penalize training errors by the loss function over the error tolerance , and C is a positive trade-off parameter that determines the degree of the empirical error in the optimization problem. Following previous researchers [41, 42], the

http://dx.doi.org/10.5772/63675

optimization was simultaneously undertaken through Lagrangian multipliers under Karush Kuhn-Tucker (KTT) conditions.

5. Results and discussion

The prime objective of using phase space reconstruction was to find a proper lag time for developing the models in this study. In order to have a comprehensive understanding of model performance, GEP models were defined by all lag times up to the optimum value determined for water demand in the CKD. The AMI calculations of the water demand in the CKD resulted in a lag time of 3 months. **Figure 3** shows that the first local minimum point occurs at 3 months,

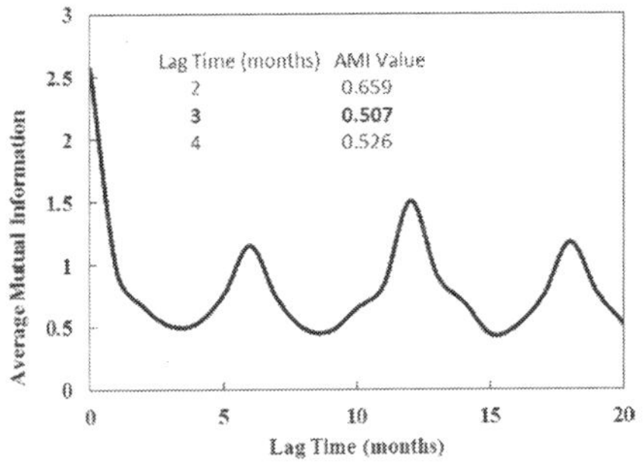

Figure 3. Average mutual information (AMI) for water demand.

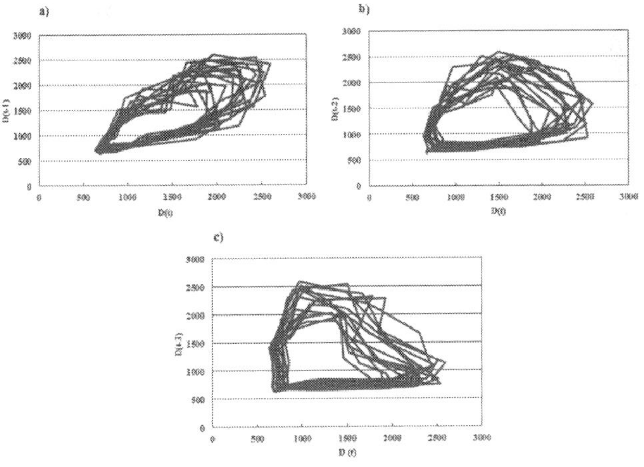

Figure 4. Phase space diagram lag times (1–3 months).

allowing the AMI an optimum lag time for phase space reconstruction (τ = 0.6591 for 2 months, τ = 0.5073 for 3 months).

Figure 4a–c shows the phase space diagrams of water demand for τ = 1, 2, and 3 months, respectively. Each figure represents the state of WDS demand at the given time. The evolution of phase space in this time series was given by reconstructing a pseudo phase space in which the demand of CKD, a nonlinear system, was considered by its self-interaction using AMI [43]. **Figure 4c** (τ =3) has a more regular pattern in comparison with the other two previous states of phase space (τ = 1, 2; **Figure 4a** and **b**, respectively), showing a lag time of 3 months to be optimum.

Prior to analysis with GEP models, a correlation table between the explanatory variables and water demand provided a better understanding of how to define the input factors (**Table 3**). The correlations were 0.92, 0.84, −0.83, 0.11, and −0.01 for *D* vs. *T*, *D* vs. HOR, *D* vs. RH, *D* vs. *P*, and *D* vs. *R*, respectively. Interestingly, water demand was highly correlated to temperature and hotel occupancy rate in CKD, showing the periodic cycle of demand due to seasonal changes. This research, however, employed all input factors in evolving the GEP models.

	D	*T*	*R*	*RH*	*P*	*HOR*
D	1.00	0.92	−0.01	−0.83	0.11	0.84
T	0.92	1.00	0.10	−0.89	0.00	0.92
R	−0.01	0.10	1.00	−0.05	−0.26	0.11
RH	−0.83	−0.89	−0.05	1.00	0.02	−0.84
P	0.11	0.00	−0.26	0.02	1.00	−0.09
HOR	0.84	0.92	0.11	−0.85	−0.09	1.00

D, demand; *P*, population; HOR, hotel occupancy factor; *T*, temperature, RH, relative humidity, and *R*, rainfall.

Table 3. Correlation between water demand and factors impacting demand.

Table 4 shows all 27 GEP models developed in the present study. Three superior models were highlighted in each category or classification of determinants. Interestingly, a lag time of 3 months outperformed other combinations in all different classifications which show the importance of using phase space construction in studying complex systems. This shows that an appropriate lag time determined by AMI can significantly improve the performance of the forecasting model. Different genetic operators were also used to understand which mathematical operations better define the nature of these determinants. The first operator {+, −, *x*} showed better performance in the first two classifications, i.e., for demand based and demand plus climatic info based categories. The second operator (OP_2) {+, −, x, x^2, x^3} outperformed other operators in (OP3) (demand + socioeconomic + climatic information) of input parameters in which socioeconomic factors were included. It is interesting that using more complex mathematical operations, as in OP_3 {+, −, x, x^2, x^3, √, e^x, log, ln} consistently reduced the quality of the models' performance. This showed that water demand forecasting could be reasonably explained by models using basic mathematical operations despite its complexity. Used to investigate the sensitivity of the models to determinant classification, the genetic operator, and lag time, the performance indices of MAE and RMSE did little to distinguish among the best performing models ($M_1D_3OP_1$, $M_2D_3OP_1$, and $M_3D_3OP_2$) in each category, i.e., MAE = 0.304, 0.3035, and 0.291, respectively, and RMSE = 0.3984, 0.3664, and 0.3660. While R^2 values showed M2 and M3 models to slightly outperform M1 models, plotting observed and predicted demand over time, as well as scatter plots of observed vs. predicted demand served to further delineate differences in performance (**Figure 5**). Comparing cumulative water demand calculated by each of the three top models to observed values showed the $M_1D_3OP_1$ and $M_3D_3OP_2$ models to be more accurate than $M_2D_3OP_1$ (**Figure 6**). In order to distinguish between $M_1D_3OP_1$ and $M_3D_3OP_2$ a plot of cumulative (observed – predicted) was plotted (**Figure 7**). This

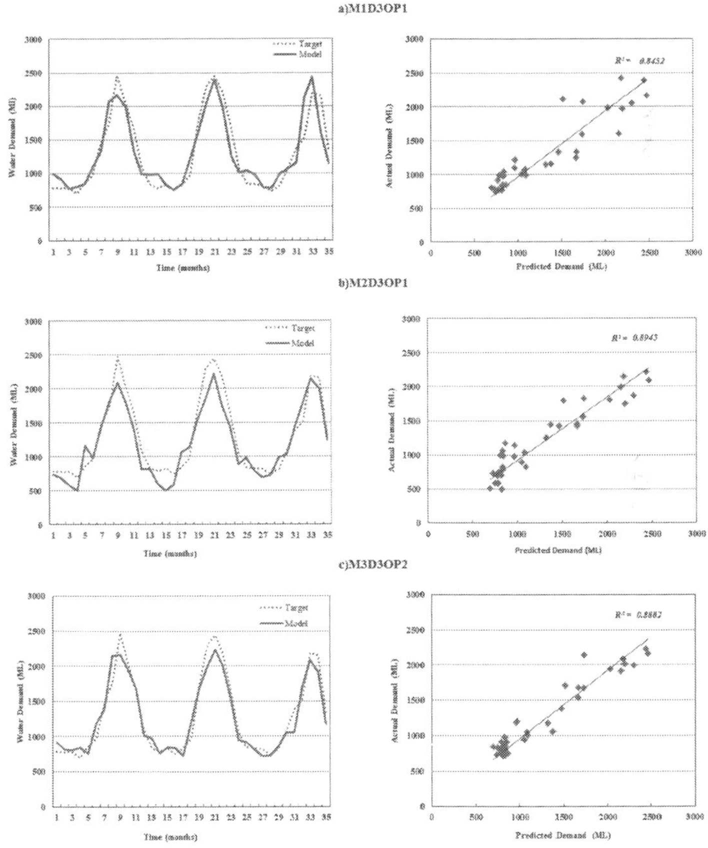

Figure 5. Observed and predicted demand over time (left), and scatter plots of observed vs. predicted demand (right) using superior GEP models: (a) $M_1D_3OP_1$; (b) $M_2D_3OP_1$; c) $M_3D_3OP_2$.

showed model $M_3D_3OP_2$ to be the best given the lesser fluctuations in errors and a consistent pattern throughout the plot's time period. This better performance may be attributable to the combination of socioeconomic factors with demand and climatic data; this might having resulted in a more consistently accurate model, which lowered the error associated compared to the other two models.

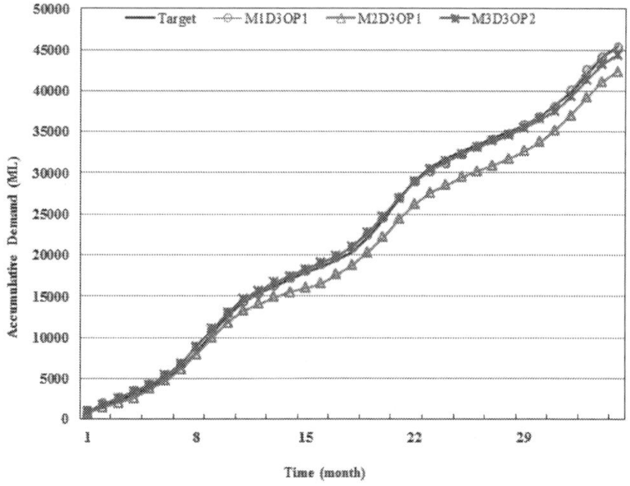

Figure 6. Cumulative demand with time.

Model ID*	Training			Testing		
	MAE	RMSE	R^2	MAE	RMSE	R^2
$M_1D_1OP_1$	0.4687	0.6974	0.6284	0.4833	0.6067	0.6343
$M_1D_1OP_2$	0.4718	0.6100	0.6252	0.4849	0.6120	0.6300
$M_1D_1OP_3$	0.4672	0.6118	0.6235	0.4800	0.6112	0.6281
$M_1D_2OP_1$	0.3552	0.4721	0.7754	0.378	0.4607	0.7892
$M_1D_2OP_2$	0.3574	0.4721	0.7756	0.3794	0.4608	0.7892
$M_1D_2OP_3$	0.3008	0.4049	0.8481	0.4188	0.5188	0.8346
$M_1D_3OP_1$	**0.3229**	**0.4317**	**0.8156**	**0.3040**	**0.3984**	**0.8452**
$M_1D_3OP_2$	0.2858	0.3641	0.8691	0.3488	0.3106	0.8452
$M_1D_3OP_3$	0.3545	0.4647	0.7849	0.3637	0.4548	0.8029
$M_2D_1OP_1$	0.3777	0.4790	0.7735	0.4529	0.5296	0.7552
$M_2D_1OP_2$	0.3955	0.4933	0.7560	0.4423	0.5169	0.7546
$M_2D_1OP_3$	0.3914	0.4893	0.7903	0.4596	0.5488	0.7643

http://dx.doi.org/10.5772/63675

Model ID*	Training			Testing		
	MAE	RMSE	R^2	MAE	RMSE	R^2
$M_2D_2OP_1$	0.2463	0.3359	0.8867	0.3015	0.3981	0.8426
$M_2D_2OP_2$	0.3236	0.4022	0.8438	0.3455	0.4176	0.8473
$M_2D_2OP_3$	0.3580	0.4450	0.8048	0.3987	0.4798	0.8077
$M_2D_3OP_1$	**0.2957**	**0.3758**	**0.8623**	**0.3035**	**0.3664**	**0.8945**
$M_2D_3OP_2$	0.3619	0.4445	0.8085	0.3893	0.4649	0.8139
$M_2D_3OP_3$	0.3033	0.4184	0.8502	0.3339	0.4562	0.8260
$M_3D_1OP_1$	0.2776	0.3810	0.8542	0.4201	0.5869	0.7087
$M_3D_1OP_2$	0.3474	0.4194	0.8237	0.4154	0.5348	0.7919
$M_3D_1OP_3$	0.2780	0.3601	0.8861	0.3933	0.5410	0.7714
$M_3D_2OP_1$	0.2875	0.3694	0.8778	0.4987	0.6332	0.6999
$M_3D_2OP_2$	0.3514	0.4543	0.8147	0.5694	0.6959	0.7027
$M_3D_2OP_3$	0.3944	0.2205	0.7827	0.5219	0.6408	0.7401
$M_3D_3OP_1$	0.3213	0.3961	0.8609	0.5624	0.6556	0.6839
$M_3D_3OP_2$	**0.2483**	**0.3230**	**0.9005**	**0.2910**	**0.3660**	**0.8882**
$M_3D_3OP_3$	0.3907	0.4801	0.7800	0.3655	0.4582	0.8236

*M_1, Demand; M_2, Demand + Climactic; M_3, Demand + Climactic + Socioeconomic; D_1, τ (lag) = 1 month; D_2, τ = 2 months; D_3, τ = 3 months; OP_1, {+, −, x}; OP_2, {+, −, x, x2, x3}; OP_3, {+, −, x, x2, x3, √, ex, log, ln}; R^2, coefficient of determination; MAE, mean absolute error; RMSE, root mean square error.

Table 4. Performance of GEP models.

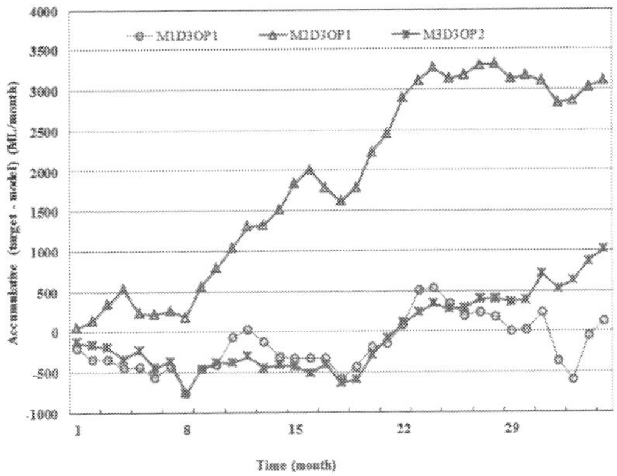

Figure 7. Cumulative (target-model) demand with time.

The superior GEP models from each classification were compared to SVM models implementing three different kernel functions (RBF, Poly, and Lin). Training and testing performance indices for the SVM models developed with each of the three kernel functions showed *Poly* kernel functions to outperform RBF and Lin functions (**Table 5**). The fact that Lin kernels performed poorly indicates that the nature of input parameters could not be considered using such functions. The M_2D_3Poly model was selected as the superior SVM model to be compared with the GEP models (**Figure 8**).

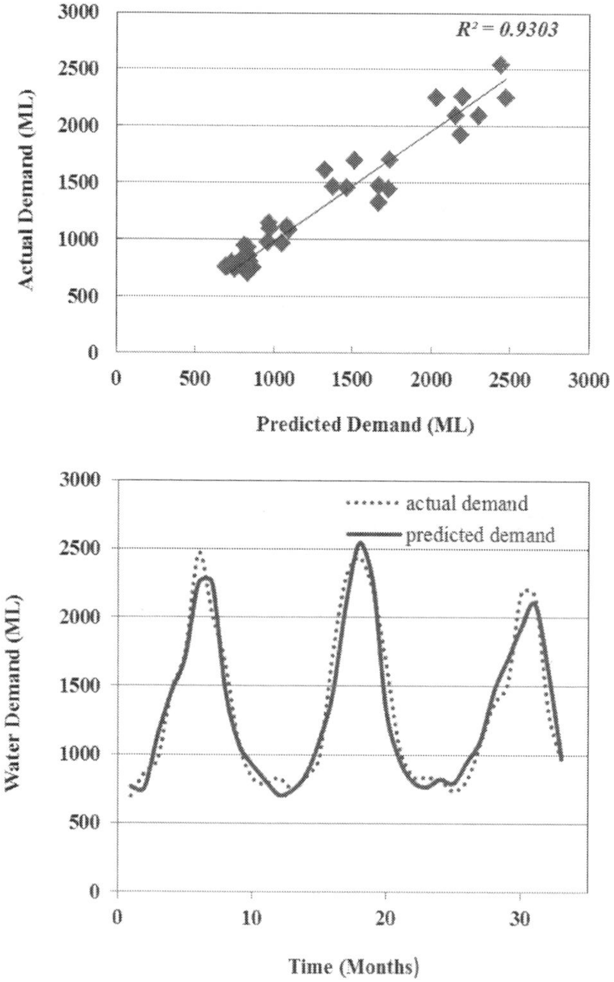

Figure 8. The best SVM model.

Model ID*	Training			Testing		
	R^2	RMSE	E	R^2	RMSE	E
M_1D_3RBF	0.9545	0.2123	0.9546	0.8397	0.4051	0.8387
M_2D_3RBF	0.9856	0.1201	0.9855	0.8701	0.3678	0.867
M_3D_3RBF	0.9416	0.2407	0.9415	0.9258	0.3014	0.9107
M_1D_3Poly	0.9308	0.2618	0.9309	0.8206	0.4278	0.8201
M_2D_3Poly	0.9372	0.2497	0.9371	0.9343	0.2593	0.9339
M_3D_3Poly	0.9428	0.239	0.9424	0.9279	0.3002	0.9114
M_1D_3Lin	0.7864	0.4602	0.7864	0.7945	0.4592	0.7927
M_2D_3Lin	0.8894	0.3311	0.8894	0.8977	0.323	0.8974
M_3D_3Lin	0.9093	0.2998	0.9004	0.9084	0.3344	0.8901

*M_1, Demand; M_2, Demand + Climactic; M_3, Demand + Climactic + Socioeconomic; D_1, τ (lag) = 1 month; D_2, τ = 2 months; D_3, τ = 3 months; RBF, Poly, Lin R^2, coefficient of determination; RMSE, root mean square error; E, Nash-Sutcliffe coefficient.

Table 5. Performance of SVM models.

6. Conclusion

In an attempt to improve model prediction accuracy, a wide range of modeling techniques has been proposed by researchers over recent years in the water demand forecasting field. The present research explored a new approach to modeling water demand, namely genetic expression programming along with phase space reconstruction. In this method, input factors are not randomly chosen as in previous studies. Instead, appropriate lag time determinations made by the AMI method defined the structure of the explanatory variables employed in the models. The outcome of this research demonstrated GEP models to be highly sensitive to classification of input factors, proper lag time, and selection of genetic operators. In general, soft computing techniques like GEP should receive more attention in forecasting behaviors of complex systems such as WDS. These models can offer valuable information to WDS operators and designers to deploy optimum determinants in their forecast models. The three best GEP models proposed in this research were compared using different performance indices, however, differentiating between them was difficult due to the similarity in statistical index values. One of three GEP models was selected due to lower cumulative error in predicting demand and less fluctuation in comparison with the other two GEP models. However, these models were slightly outperformed by a SVM model, which showed even better performance indices. This shows that both GEP and SVM can be useful techniques in water demand forecasting and can account for nonlinearity of the input parameters

Acknowledgements

The authors received financial support from the Natural Sciences and Engineering Research Council (NSERC) of Canada. The Okanagan Basin Water Board and the City of Kelowna are thanked for providing water consumption data.

Author details

Sina Shabani[1], Peyman Yousefi[2], Jan Adamowski[3] and Gholamreza Naser[4*]

*Address all correspondence to: bahman.naser@ubc.ca

1 Ph.D. Candidate, School of Engineering, University of British Columbia, Kelowna, BC, Canada

2 Ph.D. Student, School of Engineering, University of British Columbia, Kelowna, BC, Canada

3 Associate Professor, School of Engineering, McGill University, Montréal, QC, Canada

4 Assistant Professor, School of Engineering, University of British Columbia, Kelowna, BC, Canada

References

[1] Blokker M, Vloerbergh I, Buchberger S. Estimating peak water demands in hydraulic systems II – future trends. 14th Water Distribution Systems Analysis Conference, 24–27 September 2012; Adelaide. Australia: WDSA, 2012. p. 1138–1147.

[2] Ghiassi M, Zimbra D, Saidane H. Urban water demand forecasting with a dynamic artificial neural network model. Journal of Water Resources Planning and Management. 2008; 134(2):138–146. DOI: 10.1061/(ASCE)0733-9496.

[3] Troy P, Holloway D. The use of residential water consumption as an urban planning tool: a pilot study in Adelaide. Journal of Environmental Planning and Management. 2004; 47(1):97–114. DOI: 10.1080/0964056042000189826.

[4] Koo JY, Yu MJ, Kim SG, Shim MH, Koizumi A. Estimating regional water demand in Seoul, South Korea, using principal component and cluster analysis. Water Science & Technology: Water Supply. 2005;5(1):1–7.

[5] Miao, SP. A class of time series urban water demand models with non-linear climatic effects. Water Resources Research. 1990;26(2):169–178. DOI: 10.1029/WR026i002p00169.

[6] Jain A, Varshney AK, Joshi UC. Short term water demand forecast modeling at IIT Kanpur using artificial neural networks. Journal of Water Resources Management. 2001;15(5):299–321. DOI: 10.1023/A:1014415503476.

[7] Gato S, Jayasuriya N, Roberts P. Understanding urban residential end uses of water. Water Science and Technology. 2011;64(1):36–42. DOI: 10.2166/wst.2011.436.

[8] Beck L, Bernauer T. How will combined changes in water demand and climate affect water availability in the Zambezi River Basin? Global Environmental Change. 2011;21(3):1061–1072. DOI: 10.1016/j.gloenvcha.2011.04.001.

[9] Bakker M, Vreeburg JHG, van Schagen KM, Rietveld LC. A fully adaptive forecasting model for short-term drinking water demand. Environmental Modelling and Software. 2013;48:141–151. DOI: 10.1016/j.envsoft.2013.06.012.

[10] Zhou SL, McMahon TA, Walton A, Lewis J. Forecasting daily urban water demand: a case study of Melbourne. Journal of Hydrology. 2000;236(3):153–164. DOI: 10.1016/S0022-1694(00)00287-0.

[11] Mukhopadhyay A, Akber A, Al-Awadi E. Analysis of freshwater consumption patterns in the private residences of Kuwait. Urban Water. 2001;3(1–2):53–62.

[12] Dos Santos CC, Pereira Filho AJ. Water demand forecasting model for the metropolitan area of São Paulo, Brazil. Water Resources Management. 2014;28(13):4401–4414. DOI: 10.1007/s11269-014-0743-7.

[13] Zhou S, McMahon T, Walton A, Lewis J. Forecasting operational demand for an urban water supply zone. Journal of Hydrology. 2002;259(1–4):189–202. DOI: 10.1016/S0022-1694(01)00582-0.

[14] Anderson R, Miller T, Washburn M. Water savings from lawn watering restrictions during a drought year in Fort Collins, Colorado. Journal of the American Water Resources Association. 1980;16(4):642–645. DOI: 10.1111/j.1752-1688.1980.tb02443.x.

[15] Maidment D, Parzen E. Monthly water use and its relationship to climatic variables in Texas. Journal of the American Water Resources Association. 1984; 19(8):409–418.

[16] Brekke L, Larsen M, Ausburn M, Takaichi L. Suburban water demand modeling using stepwise regression. Journal of American Water Works Association. 2002;94(10):65–75.

[17] Polebitski A, Palmer R, Waddell P. Evaluating water demands under climate change and transitions in the urban environment. Journal of Water Resources Planning and Management. 2010;137(3):249-257. DOI: 10.1061/(ASCE)WR.1943-5452.0000112.

[18] Lee SJ,Wentz EA, Gober P. Space-time forecasting using soft geostatistics: a case study in forecasting municipal water demand for Phoenix, Arizona. Stochastic Environmen-

tal Research and Risk Assessment. 2010;24(2):283–295. DOI: 10.1007/s00477-009-0317-z.

[19] Alhumoud J. Freshwater consumption in Kuwait: analysis and forecasting. Journal of Water Supply Research and Technology. AQUA. 2008;57(4):279–288. DOI: 10.2166/aqua.

[20] Vemuri VR. *Artificial Neural Networks: Forecasting Time Series*. Los Alamitos, CA: IEEE Computer Society Press; 1994.

[21] Crommelynck V, Duquesne C, Mercier M, Miniussi C. Daily and hourly water consumption forecasting tools using neural networks. Proceedings of the AWWA's Annual Computer Specialty Conference.Nashville, Tennessee; 1992. p. 665–676.

[22] Bougadis J, Adamowski K, Diduch R. Short-term municipal water demand forecasting. Hydrological Processes. 2005;19(1):137–148. DOI: 10.1002/hyp.5763.

[23] Jentgen L, Kiddler H, Hill R, Conrad S. Energy management strategies use short-term water consumption forecasting to minimize cost of pumping operations. Journal of American Water Works Association. 2007;99(6):86–94.

[24] Aly A, Wanakule N. Short-term forecasting for urban water consumption. Journal of Water Resources Planning and Management. 2004;130(5):405–410. DOI: 10.1061/(ASCE)0733-9496(2004)130:5(405).

[25] Alvisi S, Franchini M, Marinelli A. A short-term, pattern based model for water-demand forecasting. Journal of Hydroinformatics. 2007;9(1):39–50. DOI: 10.2166/hydro.2006.016.

[26] Wang X, Sun Y, Song L, Mei C. An eco-environmental water demand based model for optimising water resources using hybrid genetic simulated annealing algorithms. II: Model application and results. Journal of Environmental Management. 2009;90(8): 2612–2619. DOI:10.1016/j.jenvman.2009.02.009.

[27] Wu L, Zhou H. Urban water demand forecasting based on HP filter and fuzzy neural network. Journal of Hydroinformatics. 2010;12(2):172–184. DOI: 10.2166/hydro.2009.082.

[28] Msiza IS, Nelwamondo FV, Marwala T. Water demand prediction using artificial neural networks and support vector regression. Journal of Computers. 2008;3(11):1–8.

[29] Herrera M, Torgo L, Izquierdo J, Pérez-García R. Predictive models for forecasting hourly urban water demand. Journal of Hydrology. 2010;387(1):141–150.

[30] Brentan BM, Luvizotto E, Herrera M, Izquierdo J, Pérez-García R. Hybrid regression model for near real-time urban water demand forecasting. Journal of Computational and Applied Mathematics. 2016; Article in Press.

[31] Adamowski J, Chan HF, Prasher SO, Ozga-Zielinski B, Sliusarieva A. Comparison of multiple linear and nonlinear regression, autoregressive integrated moving average,

artificial neural network, and wavelet artificial neural network methods for urban water demand forecasting in Montreal, Canada. Water Resources Research. 2012;48(1):W01528. DOI: 10.1029/2010WR009945.

[32] Shiri J, Kim S, Kisi O. Estimation of daily dew point temperature using genetic programming and neural networks approaches. Hydrology Research. 2014;45(2):165–181.

[33] Takens F. Detecting strange attractors in turbulence. In: Rand, D.A., Young, L.S. (Eds.), Lectures Notes in Mathematics, vol. 898.New York: Springer Verlag; 1981. p. 366–381.

[34] Ferreira C. Mutation, transposition, and recombination: An Analysis of the Evolutionary Dynamics. Proceedings of the 6th Joint Conference on Information Sciences (JCIS), North Carolina, USA. 2002. 614–617.

[35] Ferreira C. Gene expression programming: mathematical modeling by an artificial intelligence 2006. (Vol. 21). Springer Berlin Heidelberg.

[36] Fraser AM, Swinney HL. Independent coordinates for strange attractors from mutual information. Physical Review A. 1986;33(2):1134–1140.

[37] Holzfuss J, Mayer G. An approach to error-estimation in the application of dimension algorithms. In: Mayer-Kress, G. (Ed.), Dimensions and Entropies in Chaotic Systems.New York: Springer; 1986. p, 114–122.

[38] Hegger R, Kantza B, Schreiber T. Practical implementation of nonlinear time series methods: the TISEAN package. Chaos. 1999;9(2):413–435.

[39] Khatibi R, Sivakumar B, Ghorbani MA, Kisi O, Kocak K, Zadeh DF. Investigation chaos in river stage and discharge time series. Journal of Hydrology. 2012;414–415:108–117.

[40] Cortes C, Vapnik V. Support-vector networks. Machine Learning. 1995; 20(3):273–297.

[41] Yoon H, Jun SC, Hyun Y, Bae GO, Lee KK. A comparative study of artificial neural networks and support vector machines for predicting groundwater levels in a coastal aquifer. Journal of Hydrology. 2011;396:128–138.

[42] Jafarzadeh AA, Pal M, Servati M, FazeliFard MH, Ghorbani MA. Comparative analysis of support vector machine and artificial neural network models for soil cation exchange capacity prediction. International Journal of Environmental Science and Technology. 2016;13(1):87–96.

[43] Ghorbani MA, Khatibi R, Asadi H, Yousefi P. Inter-comparison of an evolutionary programming model of suspended sediment time-series with other local models. In: Ventura, S. (Ed.), Genetic Programming – New Approaches and Successful Applications; 2012. Rijeka, Croatia: Intech. p. 255–283. DOI: 10.5772/47801.